Nawal Gaafar (Ed.)
Christa Volkmar
Nabil El-Wakeil

Assessment of ear insects in wheat varieties: Monitoring & Managment

Nawal Gaafar (Ed.)
Christa Volkmar
Nabil El-Wakeil

Assessment of ear insects in wheat varieties: Monitoring & Managment

Südwestdeutscher Verlag für Hochschulschriften

Impressum / Imprint

Bibliografische Information der Deutschen Nationalbibliothek: Die Deutsche Nationalbibliothek verzeichnet diese Publikation in der Deutschen Nationalbibliografie; detaillierte bibliografische Daten sind im Internet über http://dnb.d-nb.de abrufbar.

Alle in diesem Buch genannten Marken und Produktnamen unterliegen warenzeichen-, marken- oder patentrechtlichem Schutz bzw. sind Warenzeichen oder eingetragene Warenzeichen der jeweiligen Inhaber. Die Wiedergabe von Marken, Produktnamen, Gebrauchsnamen, Handelsnamen, Warenbezeichnungen u.s.w. in diesem Werk berechtigt auch ohne besondere Kennzeichnung nicht zu der Annahme, dass solche Namen im Sinne der Warenzeichen- und Markenschutzgesetzgebung als frei zu betrachten wären und daher von jedermann benutzt werden dürften.

Bibliographic information published by the Deutsche Nationalbibliothek: The Deutsche Nationalbibliothek lists this publication in the Deutsche Nationalbibliografie; detailed bibliographic data are available in the Internet at http://dnb.d-nb.de.

Any brand names and product names mentioned in this book are subject to trademark, brand or patent protection and are trademarks or registered trademarks of their respective holders. The use of brand names, product names, common names, trade names, product descriptions etc. even without a particular marking in this works is in no way to be construed to mean that such names may be regarded as unrestricted in respect of trademark and brand protection legislation and could thus be used by anyone.

Coverbild / Cover image: www.ingimage.com

Verlag / Publisher:
Südwestdeutscher Verlag für Hochschulschriften
ist ein Imprint der / is a trademark of
AV Akademikerverlag GmbH & Co. KG
Heinrich-Böcking-Str. 6-8, 66121 Saarbrücken, Deutschland / Germany
Email: info@svh-verlag.de

Herstellung: siehe letzte Seite /
Printed at: see last page
ISBN: 978-3-8381-3583-0

Zugl. / Approved by: PhD Dissertation, MLU, 2010

Copyright © 2012 AV Akademikerverlag GmbH & Co. KG
Alle Rechte vorbehalten. / All rights reserved. Saarbrücken 2012

Assessment of wheat ear insects in various winter wheat varieties: Monitoring methods and management strategies in central Germany

Nawal Gaafar[1,2], Nabil El-Wakeil[1,2] , Christa Volkmar[2]

[1]Pests & Plant Protection Dept. National research Center, Dokki, Cairo, Egypt
[2]Institute of Agric. & Nutritional Sciences, Martin-Luther-University Halle-Wittenberg, Germany

Halle (Saale), Germany
2013

Abbreviations

ES: Expert system

GS: Growth stage

JKF: Julius-Kühn-Research field

L: Larvae

WBM: wheat blossom midge (this mean both orange and yellow wheat midges)

OWBM: Orange wheat blossom midge

YWBM: Yellow wheat blossom midge

TKM: Thousand Kernels Mass

TMT: Thousand Metric Tonnes

Sm1: *Sitodiplosis mosellana* gene

1. SUMMARY

Wheat (*Triticum* spp.) is a worldwide cultivated cereal crop over the world. *Sitodiplosis mosellana* (Géhin), *Contarinia tritici* (Kirby) and the thrips species *Limothrips denticornis* (Hal.) and *L. cerealium* (Hal.) are among major pests of wheat ears. The intensity of thrips and wheat blossom midges (WBM) infestations in three sites Silstedt, Halle (winter and spring) and Salzmünde (large scale field) during 2007, 2008 and 2009 were evaluated. Twelve and fifty two wheat varieties were evaluated in Silstedt and Halle central Germany, respectively. The research aimed at selecting the least infested variety to be profitably used in the forthcoming cultivation. Infestation levels were studied in flowering and milky stages (GS 65 and 73) of each variety in every single-spikelet in sample of 10 ears in the studied years.

Three methods were used to evaluate the degree of insect infestations in different wheat ear varieties. Orange wheat midge adults were monitored using pheromone traps. Wheat ears were dissected when kernels were in growth stage 73 and examined using a binocular to count the number of spikelets and damaged kernels, and to identify the insect pests present. In addition, white water traps were placed on the soil underneath each variety to collect mature larvae of wheat blossom midges (WBM) as an indicator of potential crop risk for the next year.

Silstedt site

There were significant differences in thrips and WBM densities among different varieties in both years. The highest thrips numbers were recorded as 4.5, 4.3 and 4.1 thrips/ ear part in Türkis, Global and Esket varieties, respectively, while the lowest values were recorded in 1.9 and 2.0 thrips/ ear part in Carenius and Robigus varieties. The results showed that the highest WBM infestation was observed in Türkis (5.3 larvae/ ear part); on the other hand the lowest WBM infestation was found in some resistant varieties {Brompton, Skalmeje, Robigus, Welford and Glasgow}. The damaged ears were positively correlated with the numbers of WBM among varieties.

Halle site (winter wheat)

A significant correlation between the susceptible growth stages of wheat plants (GS 47-65) and peak activity of wheat midges was analyzed. A strong correlation between adult midge trap catches and weather conditions was significantly recorded. There were significant differences in the number of thrips and WBM infesting wheat ears among varieties in both years. The highest thrips numbers were found as 66.2, 68.2 and 69.6/ ear in Akratos, Limes and Ritmo varieties, respectively in 2008 and as 14.6, 15.6 and 16.4 thrips/ ear in Michigan Amber, Elegant and Kontrast respectively in 2009. Thrips were the lowest in Thuareg (8/ ear) in 2008 and Robigus (0.8/ ear) varieties in 2009. The results showed that the highest WBM infestation level was observed in Michigan Amber 23.8 and 5 larvae/ ear in 2008 and 2009 respectively. The lowest WBM infestations were found in Türkis, Cubus, Capo, Welford and Robigus in both years. The number of damaged kernels was positively correlated with WBM among varieties. In the water traps, the highest numbers of WBM larvae were recorded in Saladin (39 larvae/ trap) in 2008 and Orlando and Glasgow varieties (55 and 53/ trap) in 2009. The lowest values were recorded in Victo, Enorm, Robigus and Welford varieties in both years.

Comparison between Halle and Silstedt

There were significant differences in thrips and WBM among varieties in both sites Halle and Silstedt. Numbers of thrips were higher in varieties Türkis and Welford in Halle and Türkis and Anthus in Silstedt, while thrips were the lowest in varieties Potenzial and Boomer in Halle and Robigus and Potenzial in Silstedt. WBM numbers were higher in varieties Tommi and Potenzial in Halle and Türkis and Dekan in Silstedt, while the least WBM numbers were observed in varieties Anthus, Welford and Robigus in both Halle and Silstedt. The ears damaged were significantly positively correlated with midge numbers among varieties and in both sites.

Halle site (spring wheat)

Susceptibility of spring wheat varieties to infestation with wheat blossom midges and thrips was studied in Central Germany in 2008 and 2009. In pheromone traps, *S. mosellana* adults started starkly and then population went down till end of the season. There was no significant difference in the number of thrips and WBM larvae between varieties in GS 65 in both years; while there were significant differences between varieties in GS 73. Thrips and wheat midge larvae numbers were significantly higher in Triso (12.3 larvae/ ear) than Sakha 93 variety (0.7/ ear) in 2008 and 3.6 and 6.5 larvae for both varieties, respectively in 2009. The damaged kernels were positively correlated with WBM numbers in both varieties. Numbers of migrated yellow midge larvae were higher in 2008 but lower in 2009 than orange midge. Yield was higher in Triso than Sakha 93 variety in both seasons.

Salzmünde site

This study aimed at studying the abundance of wheat ear insects in large scale field. A significant correlation between midge catches and weather conditions was obtained in field observations. A positive correlation between pheromone catches and ear infestation levels was recorded; it was higher in 2008 than in 2009. On the other hand, in 2007 there was no synchronization; *S. mosellana* hibernated emerged too late to coincide with the susceptible wheat growth stages. The chemical treatment applied at 2008 for highly infestation; there were significant differences in thrips numbers between treated (6.8 thrips/ ear) and untreated (27.6 thrips/ ear) after 10 days from treatment; while WBM numbers were 2.4 and 3.6/ ear in treated and untreated, respectively. Thrips and midge numbers were lower in the treated than in control. The high midge populations in water traps were recorded at growth stages 77-79 and 83 and the low populations were recorded at GS 75 and 75-77. This gives a reliable base for decision making to midges control.

Expert system

This part describes the development of an expert system designed to provide information to farmers and extension specialists as well as information for research purposes. A study to evaluate wheat ear insect pest management expert system called WMTES (Wheat midges and thrips expert system). This approach works well if the decision making process of the expert is well defined. It was conducted using results from three locations (Halle, Silstedt and Salzmünde) during three years 2007- 2009. Computer programs can help in information recovery and decision support when dealing with pest problems. These decision support tools can provide farmers with easy, rapid access to accurate information that can help them to obtain the threshold to make adequate management decisions. Plans for future field testing and expert system implementation are also discussed. Using such as expert system for controlling wheat ear insects can be successfully applied to the solution of daily problems in plant protection programs for wheat producers. Finally, the obtained results would give a good guide for choosing the proper varieties which proved highly resistant to their specific pests, as an efficient method of integrated plant protection.

Key words: Expert system, winter wheat, spring wheat, varieties, wheat midges, thrips, population densities, Halle, Silstedt, Salzmünde, insect resistance, pest managemen

2. ZUSAMMENFASSUNG

Weizen (*Triticum spp.*) ist eine weltweit angebaute Getreideart. Die Weizengallmücken *Sitodiplosis mosellana* (Géhin) und *Contarinia tritici* (Kirby) sowie die Thripsarten *Limothrips cerealium* (Halieus) und *L. denticornis* (Halieus) zählen zu den wichtigsten Schadinsekten am Getreide. In der vorliegenden Arbeit wurden in den Jahren 2007-2009 in Freilandversuchen untersucht, welche Methoden sich zur Überwachung der Schaderreger eignen. Weiterhin wurde überprüft welchen Einfluss die Koinzidenz (Zusammentreffen der Mücken mit dem sensiblen Entwicklungsstadium des Weizens) auf das Schadausmaß nimmt. Zum Einsatz kamen Pheromonfallen, die Methode der Ährenuntersuchung und Wasserschalen. Der Befall wurde in der Blüte und Milchreife des Getreides (BBCH 65 und 73) untersucht. Von jeder Prüfsorte wurden in jedem Jahr 10 Ähren auf Befall mit Weizengallmückenlarven und Thripsen untersucht. Das mitteldeutsche Trockengebiet wurde durch die Zuchtstation der RAGT 2n Silstedt (12 Sorten), das Julius-Kühn-Feld der Halle Universität (50 Winterweizen & 2 Sommerweizensorten) repräsentiert. Weiterhin stand jährlich ein Praxisschlag in Salzmünde zur Verfügung. Im Rahmen dieser Arbeit konnten Sorten ermittelt werden, die geringe Befallswerte zeigen und somit geeignet sind in der landwirtschaftlichen Praxis dazu beizutragen Ertragsverluste zu vermindern und den Einsatz von Pflanzenschutzmitteln auf ein notwendiges Maß zu beschränken.

Standort Silstedt: Es zeigten sich signifikante Unterschiede bei den Befallswerten von Weizengallmücken und Thripsen zwischen den Prüfsorten in beiden Jahren. Die Sorten Türkis, Global und Esket waren am stärksten mit Thripsen besiedelt (4.5, 4.3 und 4.1 Thripse/ Ähreteil), während die Sorten Robigus und Carenius geringe Befallswerte aufwiesen (2.0 und 1.9 Thripse/ Ähreteil). Die Ergebnisse zeigen, dass der höchste Befall mit Weizengallmückenlarven in die Sorte Türkis, (5.3 Larvae/ Ähreteil) zu beobachten war. Die niedrigsten Befallswerte zeigten die resistenten Sorten (Brompton, Skalmeje, Robigus, Welford und Glasgow). Die Resultate belegen eine positive Korrelation zwischen den Befall und den Zahlen der Gallmückenlarven pro Ähre.

Standort Halle (Winterweizen): Im Jahre 2007 traten in den Parzellen auf dem Julius-Kühn-Feld die meisten Mücken der Art S. mosellana erst nach BBCH 65 auf. Aufgrund der fehlenden Koinzidenz entstand ein signifikant geringer Schaden. Die Ergebnisse zeigen, dass auf Pflanzenschutzmaßnahmen gegen Weizengallmücken verzichtet werden kann, wenn keine Flugaktivität mittels Pheromonfalle bis zum Stadium BBCH 65 dokumentiert wird. Im Jahre 2008 zeigte eine gute Koinzidenz zwischen BBCH 47-65 und dem Nachweis aktiver Männchen von *S. mosellana* in den Pheromonfallen ein mögliches Schadrisiko an. Es bestand eine Korrelation zwischen den Pheromonfallenfängen und günstigen Wetterbedingungen. Es konnten signifikante Sortenunterschiede in beiden Kontrolljahren bei der Anzahl der Thripse und der Weizengallmückenlarven festgestellt werden. Im Jahre 2008 zeigten die Sorten Akratos, Limes und Ritmo den höchsten Befall mit Thripsen (66.2, 68.2 and 69.6/ Ähre). 2009 waren dies die Sorten Michigan Amber, Elegant und Kontrast 14.6, 15.6 and 16.4 Thripse/ Ähre. Die höchsten Befallswerte bei Weizengallmückenlarven zeigte in beiden Jahren die Sorte Michigan Amber (23.8 and 5 Larvae/ Ähre). Geringe Befallswerte konnten 2008 und 2009 den Sorten Türkis, Cubus, Capo, Welford und Robigus zugeordnet werden. Die Anzahl geschädigter Körner war positiv korreliert mit dem Weizengallmückenbefall. In den Wassserschalen wurden die höchsten Dichten an Weizengallmückenlarven 2008 bei den Sorten Saladin (39 Larvae/ Schale) gefunden, während 2009 die Sorten Orlando und Glasgow (55 und 53/ Schale) hohe Abwanderungszahlen zeigten. In beiden Kontrolljahre wiesen die Sorten Victo, Enorm, Robigus und Welford die niedrigsten Larvenwerte pro Wasserschale auf.

Standortvergleich Halle, Silstedt: Es wurden signifikante Befallsunterschiede (Thripse und Weizengallmückenlarven) bei den Prüfsorten an beiden Standorten gefunden. Die Anzahl der Thripse pro Ähre war am höchsten bei den Sorten Türkis und Welford in Halle und Türkis und Anthus in Silstedt, während die geringsten Befallswerte die Sorten Potential und Boomer in Halle und Robigus und Potenzial in Silstedt aufwiesen. Die höchsten Befallswerte an Weizengallmückenlarven pro Ähre zeigten in Halle die Sorten Tommi und Potenzial sowie Türkis und Dekan in Silstedt. Die geringsten Befallswerte lagen in Halle und Silstedt bei den Sorten Anthus, Welford und Robigus vor. An beiden Prüforten war die Anzahl befallener Ähren pro Sorte mit der Anzahl Weizengallmückenlarven pro Ähre positiv korreliert.

Standort Halle (Sommerweizen): Die Anfälligkeit von Sommerweizen gegenüber Gallmücken und Thripsen wurde in Mitteldeutschland in den Jahren 2008 und 2009 studiert. Die Ährenkontrollen bestätigten in beiden Jahren Befallsunterschiede zwischen den Prüfsorten. Zu Beginn der Milchreife waren die Ähren der Sorte Triso 2008 und 2009 signifikant stärker mit WGM Larven und Thripsen besetzt. Dabei zeigte die gelbe Weizengallmücke 2008 ein höheres Befallsniveau, 2009 war *S. mosellana* die dominierende Art. In beiden Jahren bestand eine positive Korrelation zwischen den befallenen Ähren und dem Befall mit WGM Larven. Am Halle erreichte der gut an die klimatischen Bedingungen in Mitteldeutschland angepasste

Wechselweizen Triso trotz stärkerer Schädigung (12.3 Larvae/ Ähre) ein höheres Ertragsniveau im Vergleich zur ägyptische Sorte Sakha93 (0.7 Larvae/ Ähre). Die Ursachen für die geringere Anfälligkeit der Sorte Sakha93 sind noch unbekannt.

Standort Salzmünde: Eine signifikante Korrelation zwischen *S. mosellana*-Fängen in Pheromonfallen und den Witterungsbedingungen wurde anhand 3jähriger Feldbeobachtungen festgestellt. Die Ergebnisse der Pheromonfallenfänge zeigten, dass im Jahre 2008 ein höheres Schadrisiko im Vergleich zu 2009 registriert wurde. Im Jahre 2007 bestand aufgrund einer fehlenden Synchronisation zwischen dem Flug von *S. mosellana* und dem empfindlichen Entwicklungsstadiums des Winterweizens kein Befallsrisiko. Aufgrund hoher Temperaturen im April wurde das Ährenschieben ca. 10 Tage früher als normal registriert.

Die Beurteilung des Schadpotentials anhand der Pheromonfallenfänge veranlasste 2008 eine chemische Bekämpfung auf dem kontrollierten Praxisschlag. Es konnten signifikante Befallsunterschiede bei Thripsen und Weizengallmückenlarven zwischen den behandelten (6.8 Thrips/ Ähre) und unbehandelten (27.6 Thripse/ Ähre) Kontrollparzellen nach 10 Tage erkannt werden. Mittels Wasserschalen war in der Phase der Abreife des Winterweizens (BBCH 77-79) eine gute Beurteilung der Befallssituation möglich. Die Ergebnisse zeigen, dass insbesondere beim pfluglosen Anbau von Weizen nach Weizen ein erhöhtes Befallsrisiko zu erwarten ist.

Expertensystem: Computerprogramme können in der Informationstechnologie helfen Entscheidungen im Umgang mit Schädlingen zu unterstützen. Diese Support-Tools bieten den Landwirten einen schnellen Zugang zu Detailinformationen die sie befähigen sollen, angemessene Entscheidungen im Sinne des integrierten Pflanzenschutzes in der Praxis zu treffen. Dieser Teil der Arbeit beschreibt die Entwicklung eines Expertensystems, dass Informationen aus der Forschung in Mitteldeutschland für Landwirte zur Verbesserung der Entscheidungsfindung bei der Bekämpfung von Weizengallmücken und Thripsen zur Verfügung stellt. Die Studie mit den Namen WMTES (Weizengallmücken und Thrips-Expertensystem) unterstützt Landwirte die einen schnellen Zugang zu Detailinformationen wünschen bei der Beurteilung der Gefährdungssituation ihrer Felder. In den Entscheidungsprozess werden auch Fragen über die Sortenwahl einbezogen, da die Entwicklung zeigt, dass in Zukunft auch resistente Sorten gegenüber der roten Weizengallmücke in Deutschland zur Zulassung kommen. Die Empfehlungen der Studie beruhen auf Forschungsdaten aus 3 Jahren an 3 Standorten im mitteldeutschen Trockengebiet.

3. INTRODUCTION

Cereals have been important in agriculture ever since Man started to cultivate crops. Winter wheat (*Triticum* ssp.) is one of the most important crops over the world (USDA 2007). The total agriculture area is 3.40 million hectares in the year 2009. Germany is the eighth largest producer of wheat in the world, averaging an annual production of 19,203 TMT. Of this amount, Germany consumes 15,868 TMT. Average imports for Germany are 1,856 TMT, and average exports are 5,390 TMT, making Germany the sixth largest wheat exporter in the world. Germany is the second biggest producer in Europe. It has eight of the 21 most productive regions, and they are as follows descending: Bayern, Niedersachsen, Sachsen-Anhalt (study site), Nordrhein-Westfalen, Mecklenburg-Vorpommern, Baden-Württemberg, Thüringen and Schleswig-Holstein. http://www.spectrumcommodities.com/education/commodity/statistics/wheat.html.

Wheat midges and thrips are major insects of wheat ear insect pests. The optimization of these insects control in an economically and ecologically sound manner, accomplished by the coordinated use of multiple tactics to assure stable crop production and to maintain pest damage below the economic injury level while minimizing insecticide risks to human and the environment (Freier & Rossberg 2001; Olfert *et al.* 2009; Freier *et al.* 2009).

In the last few years, there has been more focus on the orange wheat blossom midge *Sitodiplosis mosellana* (Géhin) (Diptera: Cecidomyiidae), which become a common and increasingly important pest of wheat in many places over the world. There is another wheat midge called yellow wheat blossom midge *Contarinia tritici* (Kirby) (Diptera: Cecidomyiidae), (Barnes 1928; Basedow 1971), where *S. mosellana* is found to be the most important. Both types of midges cause severe yield losses in years of high infestation as happened in Northern Hemisphere, Denmark, Northern Germany and the UK (Oakley *et al.* 1998; Mölck 2006). Larval feeding on the developing seeds causes shriveling and pre-sprouting damage and also facilitates secondary fungal attack by *Fusarium graminearium* and *Septoria nodorum* (Oakley 1994). Due to difficulties in detection of wheat midges the degree of damage to crops is often underestimated. The sex pheromone of *S. mosellana* has been identified (Gries *et al.* 2000) and a pheromone trap system for monitoring the pest was developed in the previous years (Oakley *et al.* 2005; Gaafar & Volkmar 2009; Gaafar 2010; El-Wakeil *et al.* 2010, 2013).

The wheat blossom midges can be controlled with insecticides containing pyrethroids or chlorpyrifos (Oakley *et al.* 2005). However, the use of insecticides is affecting the environment.

Both governmental restrictions and nevertheless public requirements are demanding a decrease in the use of pesticides in agriculture. In Northern Germany field experiments have shown that the timing of control is very important for the effectiveness of the insecticide used and prevention of damage to the crop, when dealing with wheat blossom midges (Schleich-Saidfar *et al.* 2007). This leads to the need of monitoring and forecasting strategies, which can provide farmers with decision making tools to evaluate the need and the timing of insecticide application.

The thrips fauna on wheat crops can cause serious damage, and methods of control are not sufficiently investigated. Lattauschke & Wetzel (1985) investigated the abundance dynamics of cereal thrips with sweep netting and also systematic sampling of plants. They found *Limothrips cerealium*, *L. denticornis* (Hal.) and *Thrips angusticeps* (Uzel) dominating in winter wheat, winter rye and winter and summer barley, as confirmed recently by Moritz (2006). Damage by thrips takes the form of white or silvery marks on leaves and ears caused by the cell contents being sucked out. Thrips feeding on the ovaries of tender wheat leads to distortion, degeneration, abortion of grains and shrivelling of the grain (Lewis 1973).

Yield losses can be expressed as reduced yield quantity or quality at harvest. Generally, the ear insects caused considerable consequences on yield as well as on the baking quality of flour (Helenius & Kurppa 1989; Bates *et al.* 1991; Lamb *et al.* 2000b). Assessing crop losses due to pests is important in making decisions about pest management based on costs and benefits and in allocating resources to the most important pests (Meers 2004).

The overall aim is to provide sustainable control of wheat ear insects by utilizing resistant and tolerant varieties in high risk situations. This topic led us to evaluate thrips and midge population and study the adaptation the potential resistant on different varieties. In the 1990s, a novel gene that conferred resistance to *S. mosellana* was discovered in Winnipeg, Canada (Lamb *et al.* 2000a). This gene was named Sm1 (McKenzie *et al.* 2002).

The main objectives of this study were:
1. Understanding and interpret pheromone trap of *S. mosellana* catches
2. Evaluating infestation level in wheat ears in the susceptible growth stages (GS 65 and 73).
3. Categorizing wheat varieties to susceptible and resistant based on thrips and midge populations.
4. Interpreting white water traps catches of two wheat midge species
5. Creating an expert system to ear insect management
6. Configuring the interaction between used methods to help the farmers and their advisors.

4. SCIENTIFIC BACKGROUND

4.1. Taxonomy, morphology, biology and economic significance of wheat midges

4.1.1. Taxonomy

Order: Diptera

Suborder: Nematocera

Family: Cecidomyiidae

Sitodiplosis mosellana (Géhin) Orange wheat blossom midge (OWBM)

Contarinia tritici (Kirby) yellow wheat blossom midge (YWBM)

4.1.2. Morphology

Table1 Morphological differences between orange and yellow wheat midges (Barnes 1928; Basedow 1971)

	S. mosellana	*C. tritici*
Adult colour	Adult is orange-red and slender with long legs which are thin and pale brown	Adult is light yellow with black head.
Length	Males are 1.49 mm long, whereas females are 1.98 mm long (Fig. 1A)	Body length reaches 1.5 mm in female, 1 mm in male.
Wings	Wings are transparent, clear, with distinct veins and 2 mm in length	Wings are opalescent with very fine hairs.
Antennae	Antennae brown	Antennae have 13-14 segments in female, 26 in male.
Ovipositor	Females are provided with a short, 1mm & stretchable ovipositor (Fig.1B)	Female bears a long ovipositor more than double her body length (Fig.1B).
Egg numbers	80 per female	30-40/ female
Larvae	Larva is reddish to orange, up to 2.5 mm long	Larva is citreous, 2 mm in length, absolutely glabrous, oblong-oval.
Infestation time	BBCH 55-59	BBCH 51-55 (Ding & Lamb 1999)

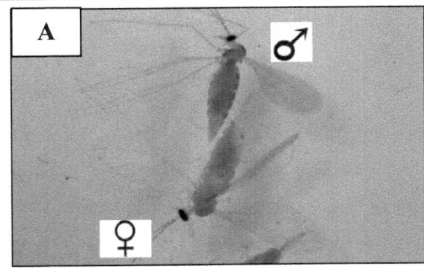 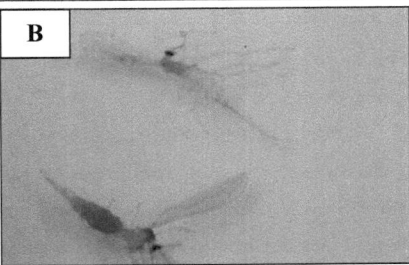

Fig. 1A Differences between male and female of orange wheat midge
Fig. 1B Differences between orange and yellow wheat midge female (photo: Gaafar)

4.1.3. Wheat midge biology
Life cycle of *S. mosellana*

S. mosellana has a very patchy spatial distribution and infestations vary from year to year depending on climatic conditions. *S. mosellana* larvae hibernate in the soil and each spring a proportion develop and pupate. It is possible for larvae to hibernate for several years if conditions are unfavourable for development of adult midges (Barnes 1956). Adult mate at the emergence site and females fly in search of a wheat crop at the ear emergence growth stage on which to lay their eggs (Oakley *et al.* 1998). In the UK, precipitation causing moist soil conditions at the end of May, followed by warm still weather in late May/early June can lead to serious *S. mosellana* outbreaks. The ovipositing female is a small insect which can remain well hidden in the crop canopy (Pivnick & Labbe 1992, 1993; Lamb *et al.* 2002). Eggs take approximately 4-10 days to hatch depending on the temperature (Smith & Lamb 2004). The larvae feed on the grain and being well hidden within the ear which is a difficult spray target. The life cycle of *S. mosellana* is univoltine (Alford 1999) Fig. 2.

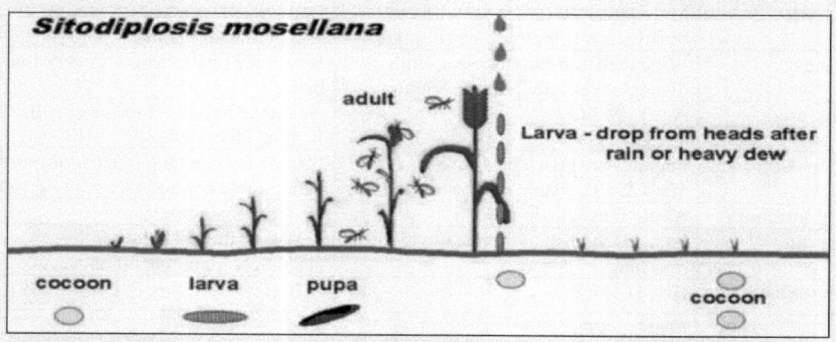

Fig. 2 Life cycle of *S. mosellana* (source Knodel 1995)

Life cycle of *C. tritici*

Adults of the yellow wheat blossom midge are broadly similar to those of the orange wheat blossom midge but tend to emerge earlier. All life cycle stages are pale yellow in colour. The life cycle is similar but with a few important differences (Barnes 1928; Basedow 1971). Female yellow wheat blossom midge adults lay eggs earlier, around GS 51-55 of the crop, and will stop laying once the floret has hardened. Each female lays a few batches of about 30 eggs, of which 4 – 15 normally survive. The eggs must hatch before pollination occurs in order that flower development can be arrested, so that the flower retains its anthers upon which the larvae then feed. If pollination succeeds the grain develops normally. Adults emerge over a shorter period than Orange wheat blossom midge and cocoons only survive up to three years in the soil (Kurppa 1989; Smith *et al.* 2004).

4.1.4. Economic significance of wheat midges

Wheat midges, *S. mosellana* and *C. tritici* are cereal specialists, for which wheat is the most attractive crop for oviposition, but in the absence of a wheat crop at a suitable growth stage, midges will fly to crops of rye, triticale, or barley, or weed grasses. Females oviposit on panicles at any stage from the onset of heading up to and including anthesis, i.e., growth stages 46–69 (Tottman & Broad 1987; Ding & Lamb 1999). Larval feeding on the developing seeds causes shrivelling and presprouting damage (Fig. 3 A&B). During a severe outbreak of orange wheat blossom midge experienced in the UK in 1993, half of the national wheat crop suffered physical damage to more than 5% of the harvested grain, 21% of crops were damaged to such an extent that a spray treatment to control wheat blossom midge would have been cost effective, had the problem been identified in time for application (Oakley, 1994). An outbreak of wheat midges occurred in winter wheat in northern Germany in 2003 (Mölck 2006) and also in UK in 2004 (Oakley *et al.* 2005) causing a high yield losses. Any

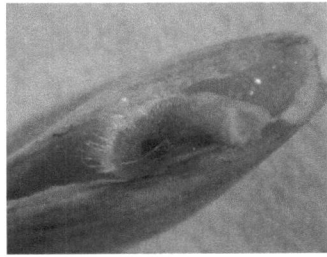

Fig. 3A *S. mosellana* damage (photo: Gaafar)

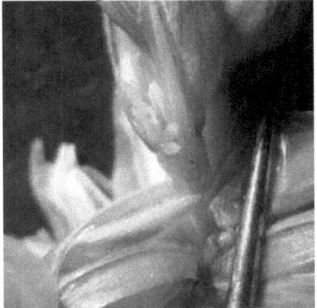

Fig. 3B *C. tritici* damage (photo: Gaafar)

insecticide application has to be applied promptly before larvae burrow in-between the lemma and palea or it will not give good control. Wheat midge management consists of early varieties, the use of insecticides, biological control and more recently the use of resistant or tolerant wheat varieties (Berzonsky *et al.* 2003). In response to concerns that further damage could occur in following years, a forecasting system was devised based on the meteorological factors known to influence the life cycle.

4.1.5. Biotic factors (wheat characteristics affecting survival of feeding larvae)

The highest WBM populations can be found in fields where wheat was grown in previous years. The Experiments have shown that the wheat growth stage at the time larvae are hatching affects larvae survival and development (Ding & Lamb 1999). Some wheat varieties produce a relatively high level of *p*-coumaric acid and/or ferulic acid in response to attack of feeding larvae. This affects the larvae, which fail to grow through the second instar. The levels of *p*-coumaric acid and/or ferulic acid increases in all wheat varieties after growth stage 59, protecting the grain against larvae attack (Ding *et al.* 2000; McKenzie *et al.* 2002). Early flowering varieties tend to come into the ear before the main emergence of *S. mosellana*. If growth stage 59 is achieved before start of oviposition, wheat plants may escape attacking of WBM (Barnes 1956; Kurppa 1989). A good flowering biology reduces the time spent in GS 55-59, which reduces the period of exposure, but may still coincide with a major oviposition event (Barnes 1956).

Resistance to *S. mosellana* is partially dominant due to expression of the *Sm1* resistance as single gene (McKenzie *et al.* 2002), that mediates an induced hypersensitive response in the surface of developing seeds where wheat midge larvae begin feeding, resulting in larval death (Ding *et al.* 2000; Lamb *et al.* 2000a). The resistance act as an antibiotic reaction that includes elevation of phenol compound levels and results in the death of the young larvae shortly after they begin feeding (Harris *et al.* 2003). The adaptations that have been studied so far in the wheat midges have been due to alleles, usually completely or incompletely recessive, at single gene in the insects (Harris *et al.* 2003). Smith *et al.* (2007) stated that the density of OWBM larvae developing on resistant wheat was lower compared to those on susceptible wheat varieties and also mentioned that small numbers of *S. mosellana* larvae matured in wheat variety was carrying the *Sm1* gene. It was also reported that OWBM infestation was associated with a reduced proportion of well-formed wheat seeds (Thomas *et al.* 2005; Ganehiarachchi & Harris 2007).

4.1.6. Abiotic factors (weather conditions)

The most critical factor identified was rainfall and its effects on diapausing midge larvae that might develop in any given season, the coincidence between emergence of adults and susceptible crop stages (Basedow 1980) and the suitability of the weather during adult activity coinciding with susceptible ear emergence growth stages for flight and oviposition (Smith & Lamb 2001; Bruce *et al.* 2007). Temperature and wind are limiting factors for *S. mosellana* flight (Pivnick & Labbe 1993).

4.2 Taxonomy, morphology, biology and economic significance of wheat thrips

4.2.1. Taxonomy

Order: Thysanoptera   Family: Thripidae

Limothrips cerealium (Hal., *L. denticornis* (Hal.) and *Thrips angusticeps* (Uzel)

4.2.2. Morphology

Thrips are 1-3 mm in size, elongated and have highly modified mouthparts which are used mainly for stabbing and sucking plant sap. The representatives of the thrips have a composite of several segments and hardened head capsule caused by the compound eyes and the broken antenna rings. The antennae of most species have seven to eight segments. The thorax consists essentially of two parts. For one of the prothorax, the connection to the head produces the cone, and the Pterothorax. On the latter, as in the word stem to be recognized Greek word πτερόν (pteron = "wings") already indicates, the wings attached, so they are present in the respective representatives (Moritz 2006). The dorsal region of the prothorax, the pronotum is trapezoidal to rectangular in thrips and has a special characteristic bristles at the edge (Moritz 2006). The ventral side of the prothorax has many areas where there are the surface membranes. The expression of the wing is within the thrips very different. These 1 and 1.2 mm long and there are about 150 to 200 of the eponymous fringes that give the wing an overall normal shape. The abdomen consists of eleven segments. This is the first segment partially located under the Thorax and the eleventh regressed considerably. The genitals are with the male at the bottom of the ninth, with the females on the bottom of the eighth segment. The tenth segment is again a distinction of the submissions (Moritz 2006).

4.2.3. Biology

The total life history from the egg to adult takes 29–35 days; egg period 10–13 days, larval period 13–17 days, pre-pupal period 2–3 days and pupal period 6–7 days. Males have short life. Thrips female overwinters in grass (Franssen & Mantel 1965; Köppä 1969) when the temperature reaches 15°C (Wetzel 1964) the females fly to winter cereals. The females feed for a period during which time the eggs mature and egg laying starts around the earrings (Köppä 1970; Larsson 2005). The number of eggs per female can be as high as 100, laid over 14–30 days. Normally, 20 eggs per leaf sheath are laid before the female moves to a new leaf sheath. The first generation is completed in winter cereals. The females migrate to summer cereals, mainly barley, where the second generation is completed (Kakol & Kucharczyk 2004).

4.2.4. Economic significance of thrips

L. cerealium feeds on winter and spring wheat. In winter cereals, thrips cause strong damage to winter wheat, whose development is related most closely to the insect life cycle (Fig. 4A& B). To a lesser degree the pest harms winter rye, barley, and other cereals. Both adults and larvae are dangerous, but the latter are usually more noxious (Meers 2004), invoking partial or complete white ear effect, drying of flag leaf, partial ear fertilization, and incomplete grain filling (Larsson 2005). During the pest outbreaks the larva density on sowing can reach 200 and more individuals per ear (Tanskii 1961). According to Tanskii the weight losses can reach 5-7% in poorly damaged grain, but 15-31% and more in strongly damaged grain. Economic injury levels for thrips (*Limothrips* spp. and *Haplothrips* spp.) in cereals in Europe have been suggested to be 5-10 thrips per ear (Seidel *et al.* 1983) and up to 30 larvae and adults per ear (Freier *et al.* 1982; Cuthbertson 1989; Parrella & Lewis 1997).

Fig. 4A Symptom of thrips adult & larvae (photo: Gaafar)

Fig. 4B Damage of thrips larvae & midge larvae (photo: Gaafar)

The upper leaf sheaths exhibit white and pale parts in the middle (Lattauschke and Wetzel 1986). Even with heavy infection in the leaf sheath, the leaves do not die (Wetzel 1964), because thrips feeding only damages the epidermis and not the leaf bundles (Mound 1971, 1997, 2005).

4.3. Expert systems

They have been developed and applied in many agriculture fields i.e. diagnose insects and diseases of various crops. Farmers across the world face problems like soil erosion, increasing cost of chemical pesticides, weather damage recovery, the need to spray, mixing and application, yield loses and pest resistance. On the other hand researchers in the field of agriculture are constantly working on new management strategies to promote farm success (Khan *et al.* 2008).

5. MATERIALS AND METHODS

5.1. Monitoring sites

Wheat orange and yellow midges and thrips were surveyed by different monitoring methods in Halle, Silstedt and Salzmünde in 2007, 2008 and 2009. These sites were chosen to cover a range of soil types and locations representative of infested areas in central Germany covered by meteorological stations. Two research fields in Halle (Latitude 51°3'N, Longitude 11°96'E) and Silstedt (Lat. 51°5'N, Long. 11°54'E) and one commercial wheat field in Salzmünde (Lat. 51°4'N, Long. 11°55'E) in the successive years were monitored (Fig.5). (http://www.mapsofworld.com/lat_long/germany-lat-long.html).

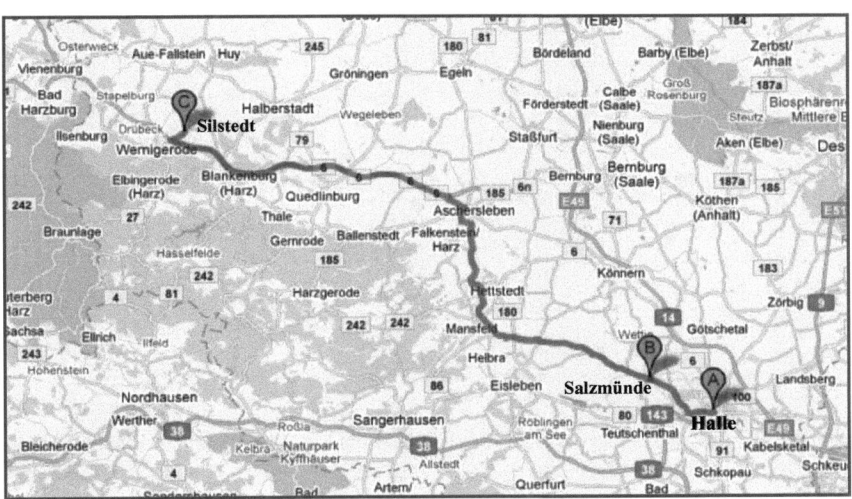

Fig. 5 Monitoring sites (Source: Google- maps)

5.1.1. Silstedt site

A. Site description

The tested varieties were cultivated in soil sandy loam in plant breeding station RAGT 2n in Silstedt, central Germany. The level of the site is 200m above sea level.

B. Weather conditions

Saxony-Anhalt is one of the temperate climate zones. It is located in the area of Transition from wet to dry region. The temperature and rainfall in Silstedt site during May to July 2008 and 2009 are presented in Fig. (6 A&B) www.wetteronline.de

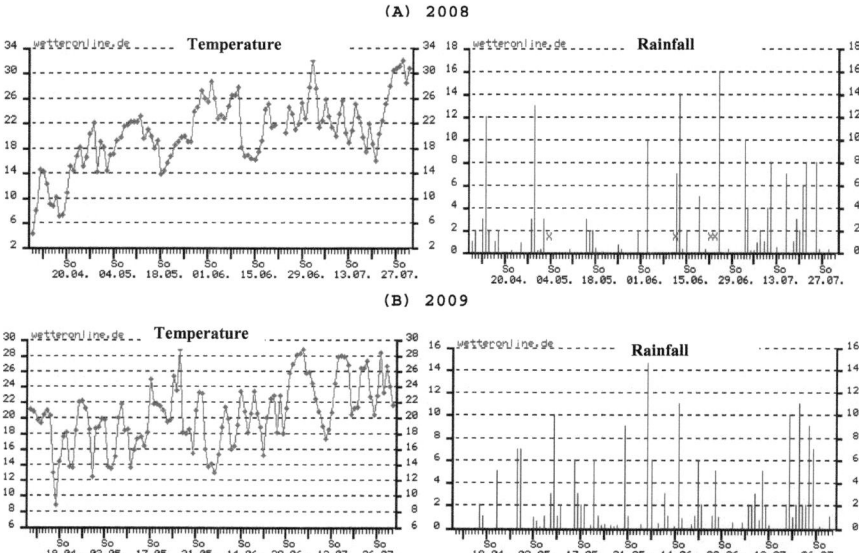

Fig. 6 Mean of daily temperature & rainfall in Silstedt from May to July 2008 (A) and 2009 (B)

C. Wheat varieties

Twelve winter wheat varieties were investigated in 2008 and 2009 (Anonymous 2009). These varieties include both German and English varieties. Five of these varieties (Glasgow, Welford, Robigus and Brompton) have proven resistance to *S. mosellana* (Ellis *et al.* 2009) as well as Skalmeje (Schliephake 2009, Personal communication). The others are susceptible varieties namely; Türkis, Tommi, Eskat, Potenzial, Global, Boomer, and Carenius.

D. Wheat field plan

The experimental area was divided into plots; 1.2m x 1.5m (1.8m^2). Two plots (replicates) were designated for each variety in a Completely Randomized Design.

E. Statistical analyses

We assumed there was not a normal distribution of the pest count data within wheat ears. The distribution of the observed data for larval counts is shown in Fig (7). In the present investigation we used a negative binomial distribution. The decision for this distribution compared with the frequently used Poisson distribution for counting data is based on a model selection with the use of the analytic criteria, AICC (Hurvich and Tsai 1989).

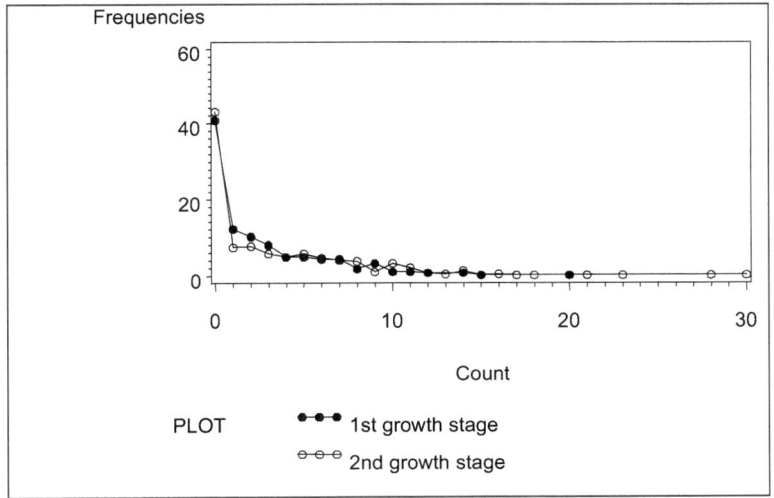

Fig. 7 Wheat midge larvae counts (observed data 2008, 1st (65) and 2nd (73) growth stages)

Accordingly we used a generalized linear model. Thus, the observed number of the i-th variety at date j in the k-th ear part of the l-th ear y_{ijkl} is assumed to be the realization of a random variable $\underline{y}_{ijkl}$ for those, the following is valid:

$$P(\underline{y}_{ijkl} = y_{ijkl}) = f_{Negbin}(y_{ijkl} \mid \mu_{ijkl}, \alpha)$$
$$(i = 1,...a; \ j = 1,..., b; \ k = 1,...c; \ l = 1,...n)$$

We have multiple measurements within the same wheat ear; i.e. counts within different ear parts (low, middle and upper). Thus, the random effect of the ear is included in the model and we have to deal with generalized linear mixed models (GLMM). Between μ_{ijkl} and η_{ijkl} we use the link function $\log(\mu_{ijkl}) = \eta_{ijkl}$, where η_{ijkl} is the so named "linear predictor".

Furthermore, in the present case we assume for the linear predictor η_{ijkl} the following model:

$$\eta_{ijkl} = \mu + \alpha_i + \beta_j + \gamma_k + (\alpha\beta)_{ij} + (\alpha\gamma)_{ik} + (\beta\gamma)_{jk} + (\alpha\beta\gamma)_{ijk} + \underline{z}_{ijl}$$
$$= \mu + \delta_{ijk} + \underline{z}_{ijl}$$

The model contains all main effects of variety, date and ear part as well as their two-way and three-way interactions. All fixed effects are summarized in the effect δ_{ijk}, which describes the effect of the i-th variety and time j and ear part k. In addition, the random ear effect $\underline{z}_{ijl}$ is included. For this effect we assume normal distribution ($\underline{z}_{ijl} \sim N(0, \sigma^2_{z(j)})$). In the course of the model selection we found a better model fit in case of time dependent ear variance estimation $\hat{\sigma}^2_{z(j)}$. Therefore we used heterogeneous variances.

The parameters of the linear predictor $\underline{\eta}_{ijkl}$ and the variance components $\sigma^2_{z(j)}$ as well as their standard errors were estimated with the use of a conditional model and the maximum likelihood method with an adaptive Gauss-Hermite quadrature. The results presented in the next paragraph are on the original scale. Using the estimates of the linear predictor and the variance components we calculate

$$\hat{\mu}_{ijk} = \exp(\hat{\mu} + \hat{\delta}_{ijk} + \frac{\hat{\sigma}^2_{z(j)}}{2}) = \exp(\hat{\eta}_{ijk} + \frac{\hat{\sigma}^2_{z(j)}}{2})$$

Derived from $\hat{\mu}_{ijk}$ we are able to make any accumulation of effects. For the estimate of the i-th variety it follows $\hat{\mu}_i = \frac{1}{b \cdot c} \sum_{j,k} \hat{\mu}_{ijk}$. The standard error of the estimates, for example $\hat{\mu}_i$, $\hat{\mu}_{i'}$ as well as their differences, was calculated by delta-method (Greene, 2003 p913ff). Statistical analysis of the differences are based on the t-value=difference/standard error (difference) and thus on the t-distribution. The calculations described above were done using the SAS-Software version 9.2 (Proc GLIMMIX, Proc IML) (SAS Institute 2009).

The relationship between numbers of thrips and midge larvae per ear was correlated with damaged kernels (shriveled, cracked, deformed kernels) in both years and among varieties using Correlation Coefficient Statistics 9 (Thomas & Maurice 2008).

5.1.2 Halle site (winter and spring wheat were cultivated here)
A. Site description

The tested winter wheat varieties were planted on the Julius-Kühn-Research field (JKF) (sandy loam soil, ca. 37 ha) in D-06112 Halle; it is an average of 113 m above sea level. The general culture conditions, such as soil type, fertilization and tillage were the same for all plots of the trial and met with local agricultural practice. There were no plant protection measures.

B. Weather conditions

Temperature and rainfall were recorded from meteorological station in Halle as shown in Fig. 8 (A&B) during May to July 2008 and 2009. The phenological observations were begun of ear emergence, flowering, milky stage and dough stage; these data were documented through twice weekly checks.

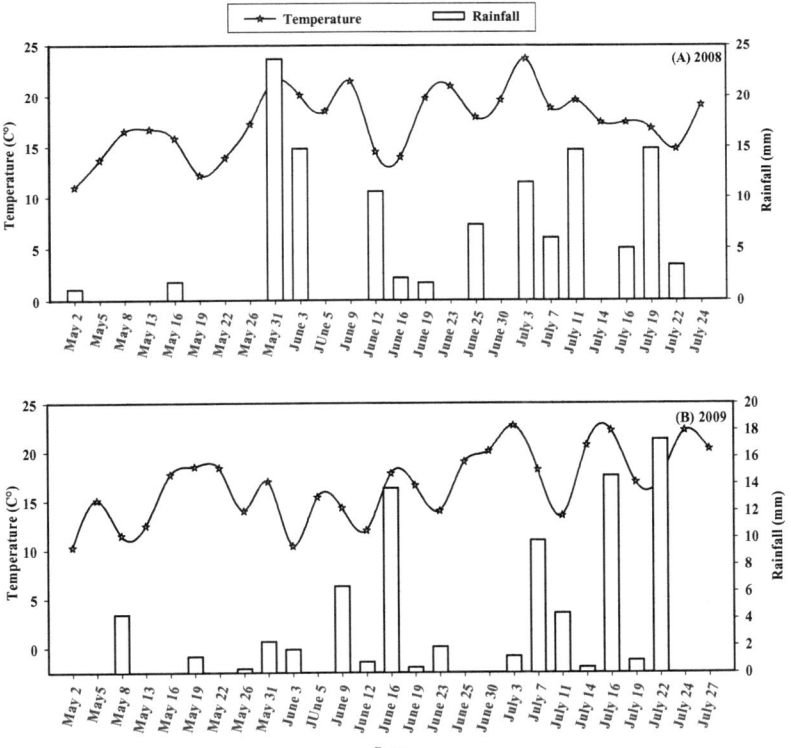

Fig. 8 Weather conditions (temperature & rainfall) in JKF-Halle from May to July 2008 (A) and 2009 (B).

C. a. Winter wheat varieties

The German Federal Office of Plant Varieties (Bundessortenamt) is responsible for variety testing and addition to the National List; their aim of the descriptive variety list is an objective description of the listed and marketed varieties in their characteristics of cultivation, resistance, quality and yield. Hundred wheat varieties were tested in 2007 and according to their infestation levels of ear insects; 50 varieties were selected for study and statistical evaluation in the upcoming test year 2008. In addition, two other varieties Glasgow and Julius were added in 2009 for consideration and discussion (Anonymous 2009). These varieties include both German and English varieties. English varieties are resistant to the orange-red wheat midges namely Robigus, Glasgow, Brompton and Welford. They have the same property sheet as German grades, but are classified differently in the valence and therefore not in accordance with the German property notes. Among the varieties are also four tribes: 15080/66, MV124/06, MV15-06, MV33-06. The studied varieties could be grouped into three categories based on their heading dates to early, middle and late as shown in Table (2).

Table 2 Order of winter wheat varieties based on their heading dates in season 2008

Variety name	Heading date	Variety name	Heading date	Variety name	Heading date
Early heading wheat varieties (22-26 May)					
Kontrast	22- May	Cardos	26- May	MV15-06	26- May
Victo	22- May	Alitis	26- May	Ludwig	26- May
MV124/06	22- May	Yukon	26- May	Skater	26- May
Marzurka	26- May	Capo	26- May	Cubus	26- May
Enorm	26- May				
Middle heading wheat varieties (28-30 May)					
Akratos	28- May	Wenga	28- May	Altos	28- May
Meteor	28- May	Transit	28- May	Terrier	28- May
Brilliant	28- May	Manager	28- May	Empire	28- May
Dekan	28- May	Magister	28- May	Glasgow	28- May
Anthus	28- May	Tommi	28- May	Julius	28- May
Limes	28- May	Michigan Amber	28- May	Toras	30- May
Leiffer	28- May	Robigus	28- May	Türkis	30- May
Potenzial	28- May	Mv33-06	28- May	Bussard	30- May
Late heading wheat varieties (02-04 June)					
Ellvis	02- June	Elegant	02- June	Welford	02- June
Solitär	02- June	Skagen	02- June	Aszita	02- June

Thuareg	02- June	**Boomer**	02- June	**Poros**	04- June
Tulsa	02- June	**Ritmo**	02- June	**Saladin**	04- June
Opus	02- June	**Hermann**	02- June	**15080/66**	04- June

C. b. spring wheat varieties

An Egyptian wheat variety (Sakha 93, known as resistant to drought) commonly cultivated in the Delta region in Egypt, and a German variety (Triso) known of high quality were sown in mid March in both years. Egyptian variety was earlier in the heading (ca. one week) than German variety.

D. a. Winter wheat field plan

The experimental area was divided into plots; 8m x 1.5m (12m^2); each variety was replicated three times and distributed in Completely Randomized Design as shown in Fig (9).

Fig. 9 Distribution of winter wheat varieties in JKF Halle

D. b. spring wheat field plan

The experimental plots were designed as a completed randomized block experiment; ten replicates were repeated in each block, with plot size measuring 1.5 x 3m. **Wheat yield:** One thousand wheat grains were weighed to count thousand kernel weight (TKM). The wheat yield was assessed in each plot to estimate the plot production in each variety. Finally, the plot production was converted to yield in kilograms/ ha as extrapolated values

E. Statistical analysis

The effect of winter or spring wheat variety on ear insect population was analysed with General Linear Model procedure with wheat variety as the fixed effect, using Statistix 9 (Thomas and Maurice 2008). This procedure computes the analysis of variance (ANOVA) for the completely randomized design. As implied by the name, the allocation of wheat varieties to the experimental units is performed completely at random. The F test assumes that the within-group variances are the same for all groups. Three homogeneity variances tests appear the ANOVA table to test this assumption (Levene 1960; Brown and Forsythe 1974; O'Brien 1981). The null hypothesis of these tests is that the wheat varieties are equal. A large F test and corresponding small p-value (< 0.05) is evidence that there are differences, Then Tukey test (post-hoc) was used to compare means of varieties.

Kernel damaged (shriveled, cracked, deformed kernels) in ears was recorded. The relationship between numbers of thrips and midge larvae per ear was correlated with damaged kernels in both years and among varieties using Statistix 9 program. The correlations procedure computes a correlation matrix for a list of varieties. Pearson or product-moment Correlations indicate the degree of linear association between varieties; then obtain values of correlation coefficients. These tests produce a value that ranges from -1 for total disagreement between rankings to 1 for total concordance. Wheat varieties were categorized to different groups using GLM Statistix 9, after ANOVA, Multiple Comparison and then test of comparison with the best (least infestation) based on Critical Value for Comparison.

5.1.3. Salzmünde site

A. Site description

The tested varieties were sown in sandy loam soil in the previous October every year in Salzmünde (Latitude 51°4'N, Longitude 11°55'E) central Germany.

B. Weather conditions:

Salzmünde is not far away from Halle (12 km); therefore same weather data were used here as presented above in Halle site.

C. Winter wheat varieties

The winter wheat varieties Tommi, Manager and Impression were chosen to be cultivated in 2007, 2008 and 2009, respectively. These varieties are commonly cultivated and with high quality properties (Anonymous 2009).

D. Wheat field plan

The tested varieties were planted in a crop rotation as winter wheat after winter wheat and the plots size was 7.5 hectares. **Chemical control:** The wheat midge management was conducted using Karate (Lambda cyhalothrin), a pyrethroid insecticide, at a rate of 0.75l/ ha (Anonymous 2008); insecticide application was sprayed on 3rd June 2008 (GS 59), and only a 4/5 of the wheat field was sprayed. Insect populations were sampled before the insecticide application, thereafter, 3, 7, 10, 15 and 20 days after treatment.

E. Statistical analysis:

The same statistical procedure which was used here, described above in Halle site

5.2. Monitoring methods

5.2.1. Survey *S. mosellana* adults using pheromone traps

Pheromone traps give reliable indication of midge emergence, onset of flight and abundance of midges throughout the season. Pheromone monitoring kits from AgriSense^{TM} (UK) were set up in winter and spring wheat fields. Two traps were set up at growth stages 45-77 (Tottman & Broad 1987) in the studied years. The pheromone traps were placed at ca. 20m from field borders and distance between two traps was 10m (Rieckmann *et al.* 2001; Petersen & Heimbach 2009). The traps were placed at the same height as the ears of the wheat plants (Fig. 10A). Trap catches were recorded twice a week (Fig. 10B). Trapped insects and debris were removed from the trap. The cards were changed depending on the density of the caught insects. Each trap consisted of a pheromone lure; Dispenser: Septa; Material: Natural rubber; Packaging: Individually Sachet Packed; Sachet Material: Foil Lined Laminate (AgriSense^{TM} 2007; Bruce *et al.* 2007).

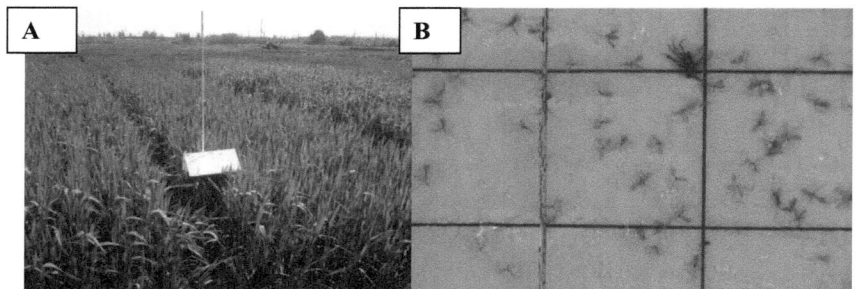

Fig. 10 Pheromone trap in winter wheat field (A) and midge adult catches (B) (photo: Gaafar)

Midge populations caught on pheromone traps were correlated with weather conditions in the different sites, mean of daily rainfall and temperature which were obtained from meteorological stations. The correlation coefficient (Pearson) was calculated using linear model by Statistix 9 program (Thomas & Maurice 2008).

5.2.2. Evaluation of thrips and midges in wheat ears

Ten ears were sampled in method of linear observation (Hübner & Wittkopf 2010) in mid of June at flowering stage (GS 65) and milky stage (GS 73) (Tottman & Broad 1987) during experimental periods. These ears were put into a bag which was then tightly sealed and taken to the laboratory where the ears were stored at -20°C. Numbers of thrips individuals (*L. denticornis*, *L. cerealium* and *T. angusticeps*), and wheat midges larvae (*S. mosellana* and *C. tritici*) were assessed using binocular. In addition, kernel damaged (shriveled, cracked, deformed kernels) in ears was recorded and presented in photo gallery. The relationship between numbers of thrips and midge larvae per ear was correlated with damaged kernels in both years and among varieties using Statistix 9 program.

5.2.3. Inspecting wheat midge larvae using white water traps

White water traps are often used to sample migrating both orange and yellow wheat midges. Larvae are caught in their migrating way from wheat ears to soil at the end of the season (Barker *et al.* 1997). The traps consisted of white plastic dishes; 12.5cm diameter (122.5 cm^2 circle surface area) and 6.5cm deep. In variety plots, one trap was placed in each plot, while in large scale field, there were two traps were placed on the ground among wheat plants near of each pheromone trap and were partly filled with water plus few drops of detergent (Fig. 11A).

The traps were observed twice a week; and the caught larvae were counted using a magnifying glass in field (Fig. 11B). Water traps were used in Silstedt site only in 2009.

 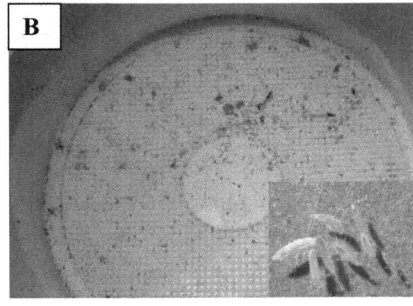

Fig. 11 White water trap (A) and midge larvae catch (B) (photo: Gaafar)

5.3. Expert system verification

Data were collected from three sites; two research fields in Halle and Silstedt and one large scale field in Salzmünde, which were selected for detailed study in 2007, 2008 and 2009. The sites were chosen to cover a range of soil types and locations representative of the infested area of central Germany, and to be cover by meteorological stations.

S. mosellana male numbers were monitored using pheromone traps and ear samples taken to assess the ultimate level of midge larvae infestation in different sites and in two growth stages; flowering (GS 65) and milky (GS 73). White water traps were used to sample the migrated wheat midge larvae to soil in the end of wheat season. For all of these sites the highest catch of male midges in pheromone traps was recorded. A correlation analysis was used to investigate the relationship between midge catches and the ultimate level of grain damaged. Levels of wheat midge infestation were relatively correlated with low/ high throughout the monitoring methods to use in the expert system.

Wheat Midges and Thrips Expert System (WMTES) is constructive computer software, giving the users a recommendation based on pheromone and water traps catches as well as infestation levels. The observations of variability in trap catch, and how it related to subsequent infestations, were very relevant when deciding how best to use the traps for wheat midge risk assessment and were used to develop a decision support model.

6. RESULTS
6.1. WINTER WHEAT VARIETIES IN SILSTEDT

Thrips were only recorded in season 2007; there were no wheat midges. These results were published in Mitt Dtsch Ges Allg Angew Ent 17 (2009) 227-230, attached in the appendix.

6.1.1. Population of *S. mosellana* adult surveyed using pheromone traps

Two peaks were recorded in 2008 (337 and 221 males/ trap) at GS 51 and 65-69, respectively (Fig. 12A); while there was only one peak in 2009 in early season (162 males/ trap) at GS 43 (Fig. 12B). Correlation between adult activity and susceptible growth stages of wheat plants was more significant in 2008 than in 2009 throughout the susceptible ear emergence growth stages. There was a strong significant correlation ($r = +0.82$) between peak pheromone trap catches and weather conditions in 2008, while a weak significant correlation ($r = +0.44$) was found in 2009 (Fig. 12 A&B).

6.1.2. Evaluation of thrips and midges in wheat ears

Three parts of wheat ear were examined for the presence of thrips and midges within two growth stages during two years. There was no significant difference in ear insect numbers between the ear parts. Therefore, the following results are presented as mean value of ear part.

6.1.2.1. 2008
Thrips (larvae and adults)

Significant differences were found ($P = 0.011$) in the number of total thrips per ear part among varieties. Esket and Türkis varieties had the highest numbers of thrips 5.7 and 6.2/ ear part, respectively, while moderate values (2.5 & 2.6 individuals/ ear part) were recorded in varieties of Welford and Carenius. The lowest number of total thrips was found in Glasgow and Robigus varieties (2.1 & 2.2 individuals / ear part), respectively (Fig. 13A).

There was a significant difference between Türkis and all varieties except Esket ($P = 0.91$). There were significant differences between Esket and either Boomer ($P = 0.038$), Potenzial ($P = 0.007$), Tommi ($P = 0.005$), Global ($P = 0.007$), Carenius ($P = 0.0008$), Welford ($P = 0.001$), Skalmeje ($P = 0.001$), Brompton ($P = 0.0003$), Robigus ($P = 0.001$) or Glasgow ($P = 0.001$). Also, significant differences were found between Boomer and either Robigus ($P = 0.040$) or Glasgow ($P = 0.044$). Moreover significant differences were also recorded among Potenzial, Tommi, Global, Carenius, Welford & Skalmeje and Robigus or Glasgow as well (Fig. 13A).

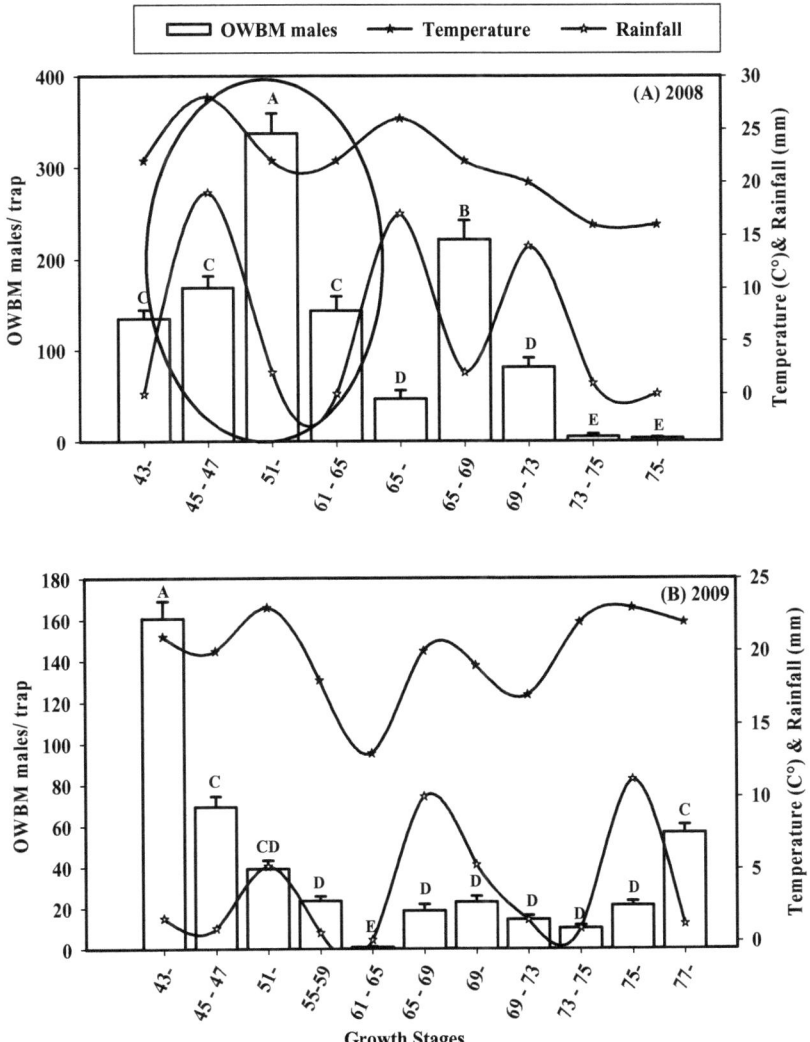

Fig. 12 Mean ± SE of *S. mosellana* male catches in pheromone traps and their relation to temperature and rainfall in 2008 (A) and 2009 (B). Oval refers to coincidence of adult activity and susceptible growth stages. Different letters indicate significant differences

Wheat midge larvae

Significant differences ($P= 0.001$) were also found between resistant and susceptible varieties (Fig 13B). There were significant differences ($P= 0.014$) among susceptible varieties, whereas the highest midge larval population (10.2 larvae/ ear part) has been found in Türkis as compared to other varieties, while the moderate varieties were reported in Carenius and Esket with value of 3.65 individuals/ ear part. The least numbers of larvae were recorded in resistant varieties namely Brompton, Skalmeje, Robigus, Welford and Glasgow (0.0, 0.1, 0.3, 0.6 and 1.2 individuals/ ear part), respectively, there was no significant difference among resistant varieties (Fig 13B). There were four groups that have significantly different numbers of larvae. Most susceptible was Türkis. Tommi, Potenzial, Global and Boomer were less susceptible as Türkis, but with significantly higher larvae than Esket and Carenius, as well resistant varieties, Glasgow, Welford, Robigus, Skalmeje and Brompton.

There was a significant difference between Türkis and all other varieties without exception. Significant differences were recorded among Tommi and all varieties except Boomer ($P= 0.051$), Global ($P= 0.570$), and Potenzial ($P= 0.777$). There were significant differences among Potenzial and other varieties except Boomer ($P= 0.115$) and Global ($P= 0.810$). Significant differences were obtained among Global and other varieties except Boomer ($P= 0.153$). Significant differences were recorded among Boomer and all varieties except Global, Potenzial and Tommi. There were significant differences among Esket and all varieties except Carenius ($P= 0.871$). Significant differences were found among Glasgow, Welford, Robigus, Skalmeje & Brompton and susceptible varieties; while there was no significant difference within these resistant varieties (Fig 13B).

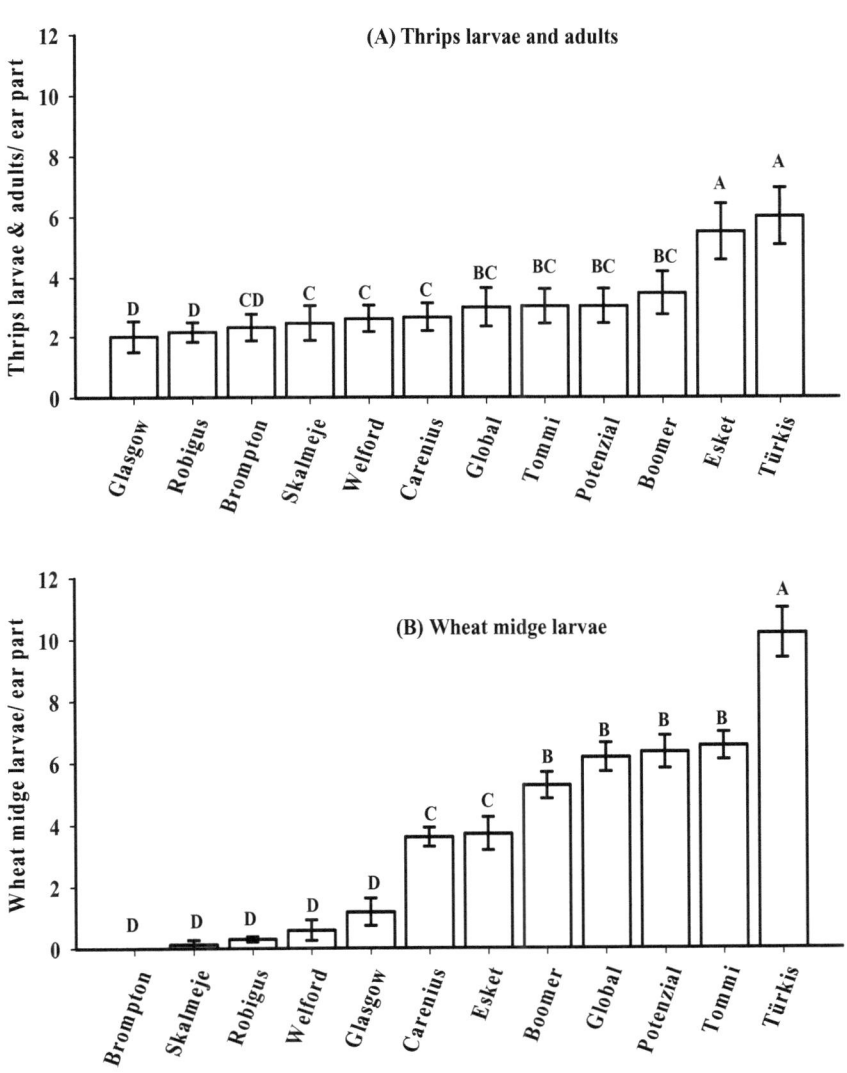

Fig. 13 Population of thrips larvae & adults (A) and wheat midge larvae (B) in different winter wheat (resistant & susceptible) varieties in Silstedt 2008. Different letters indicate significant differences.

6.1.2.2. 2009
Thrips (larvae and adults)

Significant differences were found ($P= 0.029$) in the number of total thrips per ear part among varieties. Global variety showed the highest numbers of thrips (5.6 individuals/ ear part), while moderate numbers (2.4 and 2.7 individuals/ ear part) were recorded in varieties of Esket and Boomer. The lowest number of total thrips was found in Carenius and Brompton varieties (1.3 and 1.6 individuals/ ear part), respectively (Fig. 14A).

There were significant differences between Global and all varieties as well as among Skalmeje and either Esket ($P= 0.001$), Tommi ($P= 0.008$), Potenzial ($P= 0.001$), Robigus ($P= 0.001$), Brompton ($P= 0.003$) or Carenius ($P= 0.001$) (Fig. 14A). Also, significant differences have been found among Glasgow, Esket ($P= 0.49$), Tommi ($P= 0.048$), Potenzial ($P= 0.022$), Robigus ($P= 0.016$), Brompton ($P= 0.006$) and Carenius ($P= 0.001$). Significant differences were also obtained among Türkis, Welford & Boomer and the following varieties (Brompton and Carenius) and also between Esket and Carenius ($P= 0.003$) (Fig. 14A).

Wheat midge larvae

In 2009, the resistant varieties (with exception of Skalmeje) were significantly separated from the other varieties. Significant differences ($P= 0.009$) were found between resistant and susceptible varieties (Fig 14B). There were significant differences ($P= 0.034$) among susceptible varieties, whereas Potenzial showed the highest larval population (0.7 larvae/ ear part) compared to other varieties, while the moderate varieties were Boomer and Global with values of 0.2 and 0.3 individuals/ ear part, respectively. The least WBM values were recorded in resistant varieties (Glasgow, Robigus, Brompton, Welford, and Skalmeje) ranged from 0.01 to 0.2 individuals/ ear part (Fig 14B).

Significant differences were found between: (i) Potenzial and other varieties [except Carenius ($P= 0.758$) and Tommi ($P= 0.769$)]; (ii) Tommi or Carenius and all other varieties [except Türkis and Esket]; (iii) Türkis and rest of the varieties [except Carenius, Tommi and Esket], Esket and all other varieties [except Global, Boomer and Skalmeje]; (iv) Global and Boomer and other varieties except [Esket and Skalmeje] and Skalmeje, Welford, Brompton, Robigus and Glasgow and susceptible varieties were found significantly different (Fig 14B).

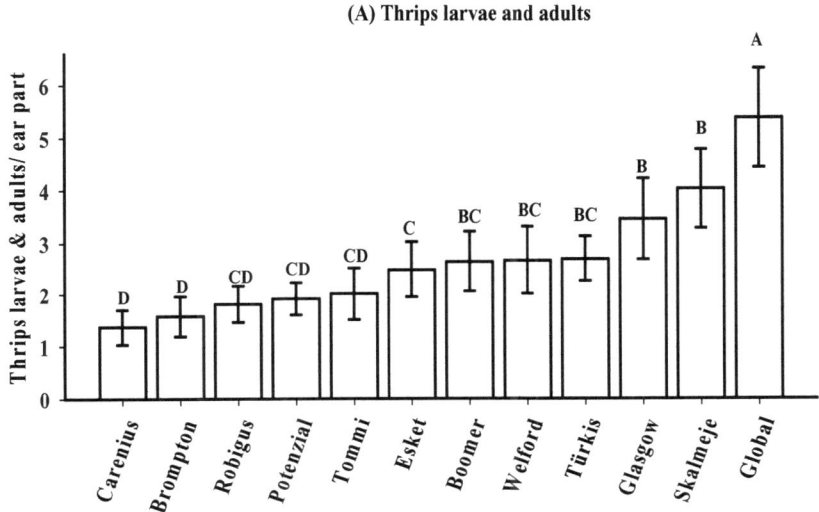

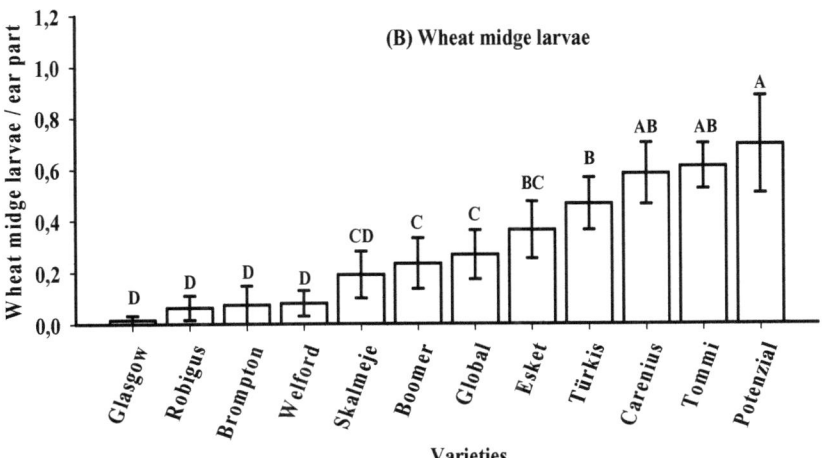

Fig. 14 Population of thrips larvae and adults (A) and wheat midge larvae (B) in different winter wheat (resistant & susceptible) varirties in Silstedt in 2009. Different letters indicate significant differences.

6.1. 2. 3. Pooled data of two years 2008 & 2009
Thrips (larvae and adults)

Generally, infestation of thrips was higher in 2008 compared to those in 2009. This may be due to the environmental conditions between the two years. There were significant differences ($P= 0.001$) between the two years in the following varieties (Carenius, Esket, Potenzial, Global and Türkis), while there was no significant difference ($P= 0.06$) in varieties of Boomer, Tommi, Brompton, Glasgow, Robigus, Skalmeje and Welford between both seasons. Significant differences were found ($P= 0.022$) in the number of total thrips per ear part among varieties. Türkis, Global and Esket varieties had the highest numbers of thrips 4.5, 4.3 and 4.1/ ear part, respectively, while moderate ones (2.6 and 2.8 individuals/ ear part) were recorded in varieties of Welford and Glasgow. The lowest number of total thrips was found in Carenius and Robigus varieties (1.9 and 2.0 individuals/ ear part), respectively (Fig. 15A).

There were significant differences between Türkis and all varieties [except Global ($P= 0.999$)]. Significant difference was obtained among Global and Glasgow ($P= 0.008$), Welford ($P= 0.005$), Tommi ($P= 0.001$), Potenzial ($P= 0.005$), Carenius ($P= 0.001$), Robigus ($P= 0.001$) and Brompton ($P= 0.001$). Also, there were significant differences between Esket and each of Glasgow, Welford, Tommi, Potenzial, Carenius, Robigus and Brompton. Significant differences were recorded between Skalmeje & Boomer against Carenius ($P= 0.001$), Robigus ($P= 0.012$) and Brompton ($P= 0.002$) and also among Glasgow, Welford, Tommi & Potenzial against Carenius, Robigus and Brompton (Fig. 15A).

Wheat midge larvae

The proportion of ears damaged with wheat midges also differed significantly ($P= 0.0025$) between the two years. In general, wheat midge infestation was lower in 2009 compared to those in 2008. There were significant differences ($P= 0.001$) in both years among all varieties except in the varieties of Brompton, Skalmeje and Welford. Significant differences ($P= 0.006$) were found among resistant and susceptible varieties (Fig 15B). There were significant differences ($P= 0.019$) among susceptible varieties, whereas Türkis had the highest midge larvae population (5.3 larvae/ ear part) compared to other varieties., The moderate varieties were Esket and Carenius with value of 2.1 individuals/ ear part. The least WBM values were recorded in resistant varieties namely Brompton, Skalmeje, Robigus, Welford and Glasgow (0.1, 0.2, 0.2, 0.3

and 0.6 individuals/ ear part), respectively, there was no significant difference among resistant varieties (Fig 15B).

There was a significant difference among Türkis and all varieties. Significant differences were recorded among Tommi, Potenzial & Global and all varieties except Boomer (P=0.051). Significant differences were evaluated between Boomer and all varieties [except Carenius and Esket], also between Carenius and all varieties (except Esket ($P= 0.862$)). Significant differences were found among resistant varieties (Glasgow, Welford Robigus, Skalmeje and Brompton) and susceptible varieties; while differences between resistant varieties were not significant (Fig 15B).

The number of midge larvae per ear was significantly positively correlated ($r= +0.99$) with the percentage of infested ears. There was no significant correlation between total thrips and damaged kernels ($r= +0.085$) (Table 3).

Table 3 Correlation coefficient between ear insects (thrips & wheat midges) and damaged kernels in Silstedt 2008 and 2009

Studied years	Total thrips	Wheat midge larvae
2008	+0.159	+0.99 *
2009	+0.012	+0.99 *
2008& 2009	+0.085	+0.99 *

* Significant differences are at 0.05 levels.

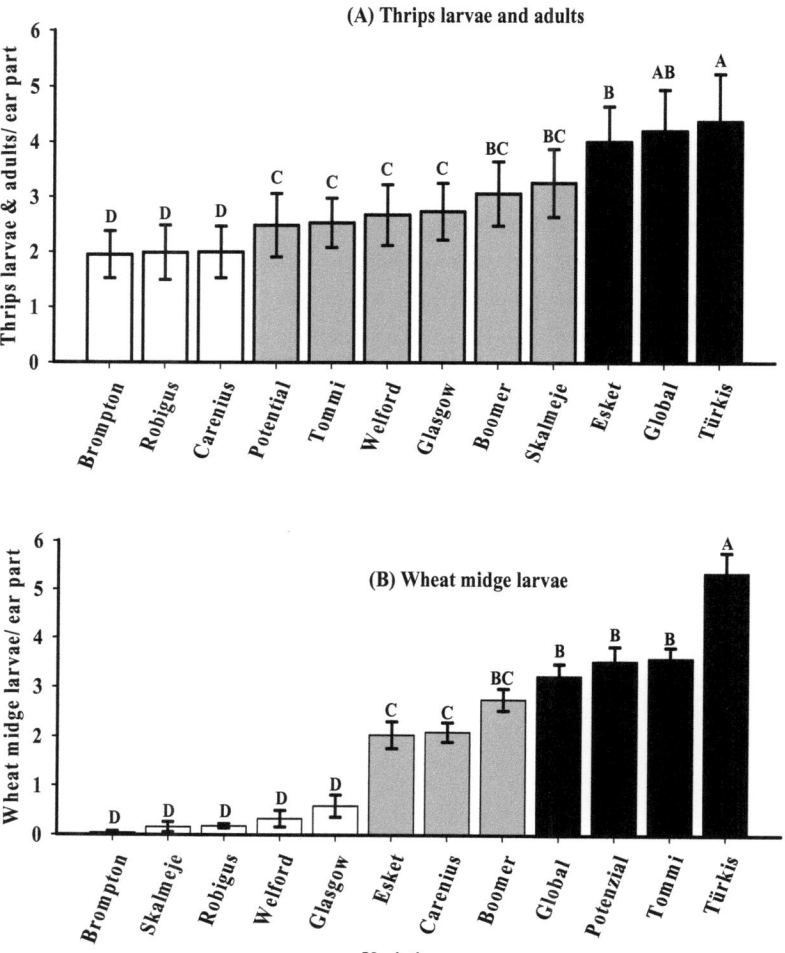

Fig. 15 Population of thrips larvae and adults (A) and wheat midge larvae (B) in different winter wheat (resistant & susceptible) varieties in Silstedt in 2008 & 2009.
Different colors indicate different infestation grades. Different letters indicate significant differences.

6.1.3. Wheat midge larvae population inspected using water traps in 2009
S. mosellana

There were significant differences ($P= 0.0013$) in the number of orange midge larvae *S. mosellana* per trap among varieties. Glasgow and Skalmeje varieties had the highest number of *S. mosellana* larvae 144 and 117 larvae/ water trap (although they are resistant varieties to OWBM) compared to other varieties. *S. mosellana* larvae in these varieties were migrated very early GS 71 in the first larval instars reached to 23 larvae/ trap in GS 75 in both varieties. While the moderate variety was Esket with value of 51/ trap. The least *S. mosellana* value was recorded in Global (25/trap) (Fig. 16A).

C. tritici

There was significant difference ($P= 0.033$) in the number of yellow wheat midge, *C. tritici* larvae per trap among varieties. Boomer and Robigus had the highest numbers of *C. tritici* larvae 3.5 and 3 larvae/trap, respectively, while middle *C. tritici* numbers were recorded in Tommi and Skalmeje varieties with values of 4/ trap. The least *C. tritici* value was recorded in varieties of Türkis, Welford, Brompton and Glasgow with value of 0.5/ trap/ variety (Fig 16B).

Wheat blossom midge larvae caught on different wheat growth stages

Total of wheat midge larvae from all the studied varieties were summed during different growth stages (GS 71-89). This population density was significantly higher ($P= 0.021$) on some growth stages (83-89) than other growth stages (71-77). The low numbers ranged from 21 to 180 midge larvae/ trap which recorded in GS 71 to 77, while the high numbers ranged from 236 to 361 midge larvae / trap which recorded in GS 83 to 89. Analyses of the cumulative data using ANOVA to compare total WBM larvae numbers showed that there was a significant difference ($P= 0.024$). The last WBM larvae were caught on growth stage 89 (Fig 17).

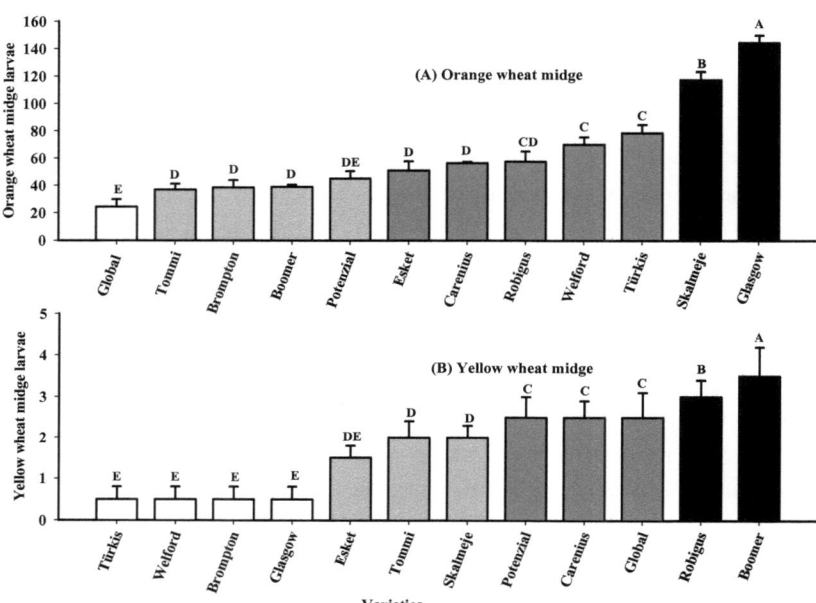

Fig. 16 Total of caught of orange midge larvae (A) and yellow midge larvae (B) per variety by white water traps in some winter wheat varieties in Silstedt 2009. Different letters indicate significant differences.

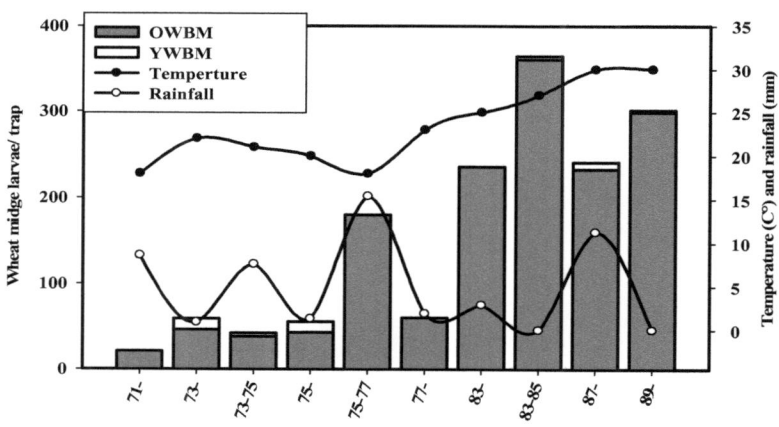

Fig. 17 Total of orange and yellow wheat midge larvae catches by water traps and their relation to temperature and rainfall during winter wheat season in Silstedt 2009

6.1.4. Categorizing of wheat varieties based on ear insect infestation

The results indicated that winter wheat varieties could be grouped into three categories based on thrips and wheat midges' infestation as shown in Fig (15) in different colors among these groups in Table (4). The results indicated that the highest number of ear insects (thrips and midges) was found in varieties of Türkis and Global, while the lowest values were recorded in Brompton and Robigus varieties.

Table 4 Infestation grades of wheat varieties based on population of thrips and wheat midge larvae in Silstedt 2008 & 2009. Different letters indicate significant differences.

Grades	Total thrips/ ear part		Grades	Wheat midge larvae/ ear part	
	Varieties	Mean		Varieties	Mean
Low	Brompton	1.948 D	Low	Brompton	0.036 D
	Robigus	1.987 D		Skalmeje	0.159 D
	Carenius	1.994 D		Robigus	0.179 D
Moderate	Potenzial	2.483 C		Welford	0.329 D
	Tommi	2.530 C		Glasgow	0.590 D
	Welford	2.673 C	Moderate	Esket	2.034 C
	Glasgow	2.736 C		Carenius	2.093 C
	Boomer	3.062 BC		Boomer	2.751 BC
High	Skalmeje	3.257 BC	High	Global	3.220 B
	Esket	4.002 B		Potenzial	3.520 B
	Global	4.197 AB		Tommi	3.575 B
	Türkis	4.367 A		Türkis	5.332 A

6.2. WINTER WHEAT VARIETIES IN HALLE

Ear insect populations were very low in season of 2007; although three growth stages, heading (59), flowering (65) and milky (73) in 100 wheat varieties were investigated. Mean temperature in April was 18.2°C and rainfall 0.1mm. Therefore there was no coincidence between wheat midge activity and susceptible growth stages which matured early. Therefore, pheromone trap results were only presented here as example to show this situation (Fig. 18). More information is published in Mitt Dtsch Ges Allg Angew Ent 17 (2009) 221-225, attached in the appendix.

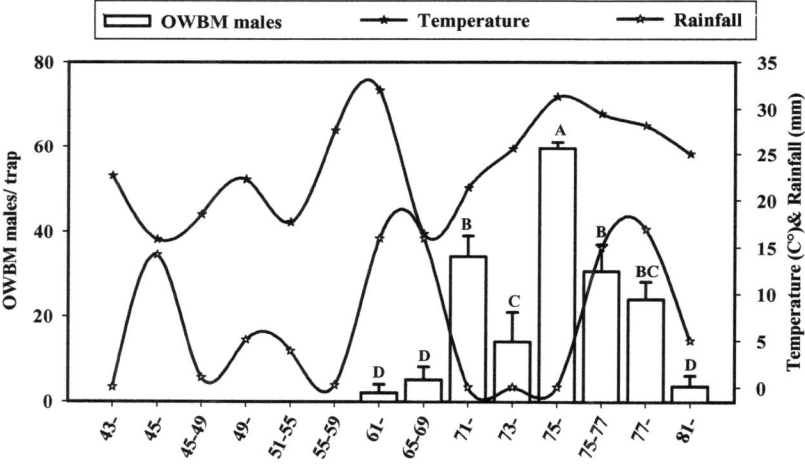

Fig. 18 Mean *S. mosellana* male catches in pheromone traps and their relation to temperature and rainfall in Halle 2007. Different letters indicate significant differences

6.2.1. Population of *S. mosellana* adult surveyed using pheromone traps

In Halle 2008, three peaks were recorded (164, 140 and 128 midges/ trap) at GS 51-59, 55-59 and 61-65, respectively (Fig. 19A). There were also there peaks in 2009 (47, 60 and 60 midges/ trap) at GS 59-61, 61-65 and 65-69, respectively (Fig. 19B). Generally wheat midge adult populations were higher in 2008 than 2009. Correlation between adult activity and susceptible growth stages of host plants was more significant in 2008, and allowed a higher infestation to develop from lower population levels.

Correlation between midges caught and weather conditions

Large variations in numbers of midges caught in the pheromone traps (approximately threefold) and in timing of peak catches were found between years. Catches were more correlated with susceptible growth stages in 2008 than 2009. There was a strong correlation between peak pheromone trap catches and weather conditions ($r = 0.82$, $r = 0.67$) in both years, respectively as shown in Fig. 19 (A&B).

6.2.2. Evaluation of thrips and midges in wheat ears

6.2.2.1. 2008

Total thrips (larvae and adults)

There was a significant difference ($P= 0.0001$) in the number of total thrips per ear among varieties. Akratos, Limes and Ritmo varieties had the highest numbers of total thrips 66.2, 68.2 and 69.6/ear, respectively, while moderate population levels of 26.8, 27.0 and 27.8/ear were recorded in varieties of Enorm, Cubus and Tommi, respectively. The lowest number of thrips was found in Thuareg variety (8/ear) (Fig. 20A).

Wheat midge larvae

There was also a significant difference ($P= 0.0003$) in the number of both wheat midge larvae per ear among varieties. Michigan Amber, Skater and Terrier had the highest numbers of WBM larvae 23.8, 13.8 and 9.8 larvae/ ear, respectively, while the moderate varieties were Bussard, Potenzial and Skagen with value of 1.2/ear. The least wheat midge larvae were recorded in Boomer (0.2/ear). Midge larvae were not found in some varieties namely, Türkis, Anthus, Transit, Enorm, Solitär, Magister, Marzurka, Victo, and Welford, the latter is a variety that is resistant to wheat midge (Fig 20B).

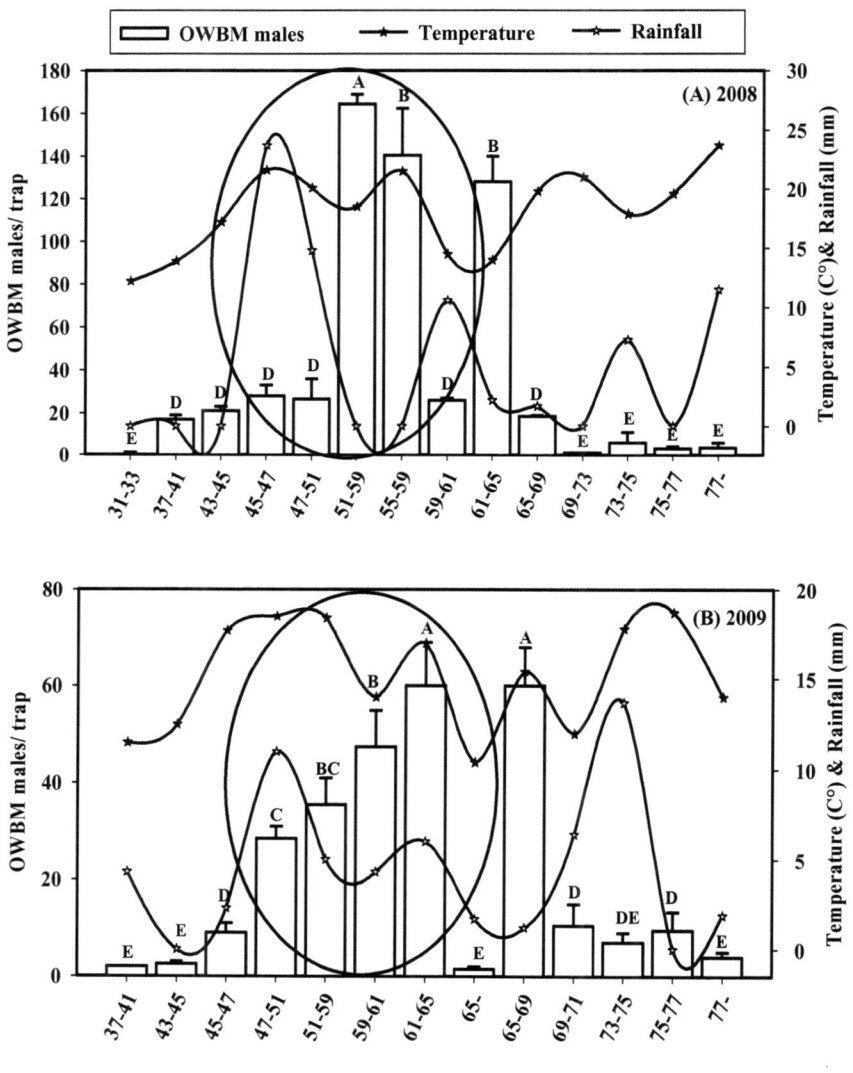

Growth stages
Fig. 19 Mean ± SE of *S. mosellana* male catches in pheromone traps and their relation to temperature and rainfall in 2008 (A) and 2009 (B). Ovals refer to coincidence of adult activity and susceptible growth stages. Different letters indicate significant differences

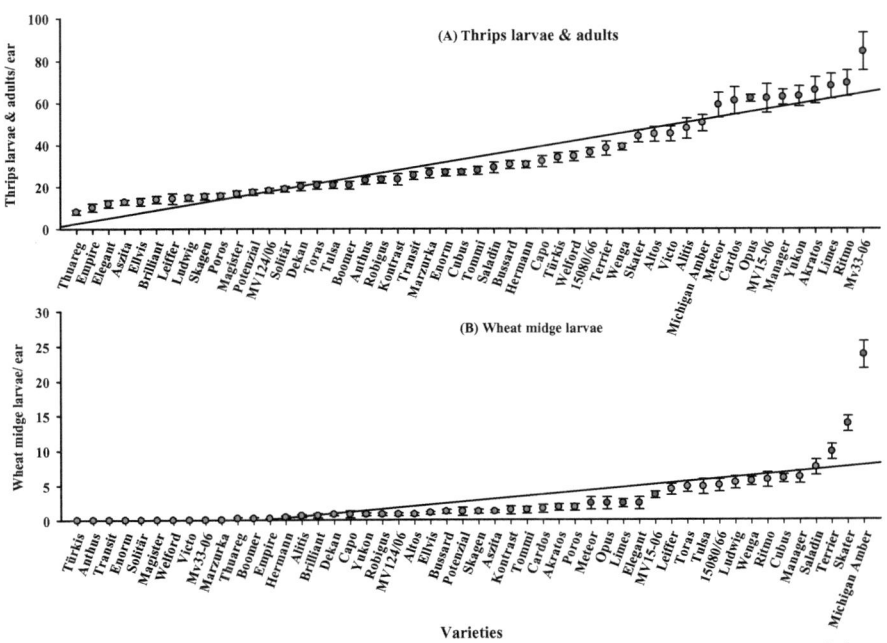

Fig. 20 Mean ± SE of thrips larvae & adults (A) and wheat midges (B) per wheat ear in 50 winter wheat varieties (ordered based on insect infestation) in milky stage (GS 73) in 2008

There were significant differences ($P= 0.0005$) in the number of kernels damaged by thrips and both wheat midges among varieties. The highest number of damaged kernels was found in Skater, Terrier and Michigan Amber varieties at 3.6, 5.4 and 7.0/ear respectively, while Bussard, Capo and Yukon had moderate infestations (0.6/ear). There were number of varieties with no damaged kernels namely Türkis, Anthus, Transit, Enorm, Solitär, Magister, Welford, Victo, Empire and Marzurka (Fig 21). There was positive correlation between WBM infestation and damaged kernels ($r = +0.66$), while there was no significant correlation between thrips and damaged kernels ($r = +0.11$) in total thrips (Table 5).

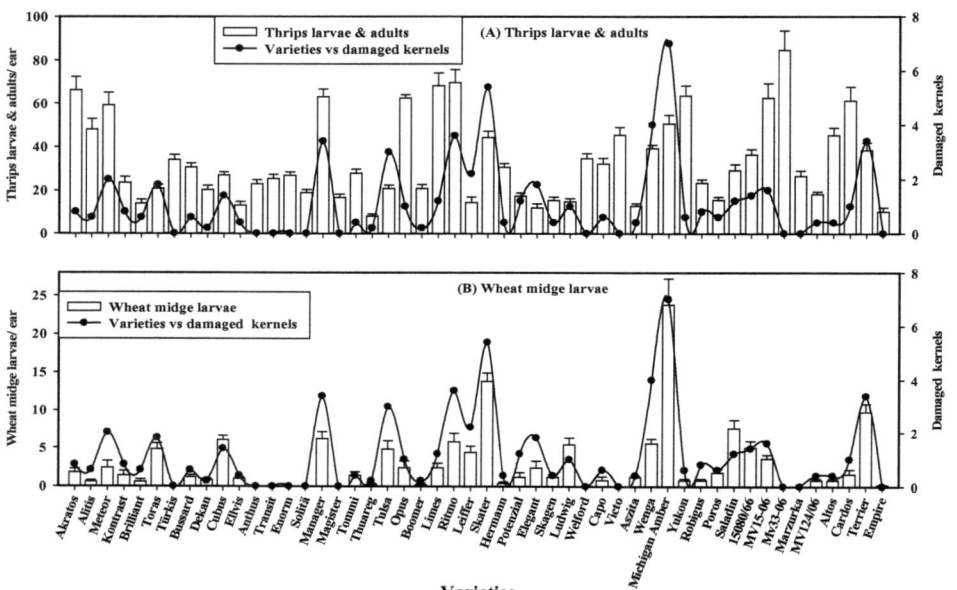

Fig. 21 Mean ± SE of thrips larvae & adults (A) and wheat midges (B) per wheat ear and their correlation to damaged kernels in 50 winter wheat varieties in milky stage (GS 73) in 2008

Table 5 Correlation coefficient between ear insects (thrips and wheat midges) and damaged kernels in Halle 2008 and 2009

Studied years	Total thrips	Wheat midges
2008	+0.11	+0.66 *
2009	+0.17	+0.76 *
Pooled 2008&2009	+0.15	+0.75 *

* Significant differences are at 0.05 levels.

6.2.2.2. 2009

Total thrips (larvae and adults)

There was a significant difference ($P = 0.0003$) in the number of total thrips per ear among varieties. Michigan Amber, Elegant and Kontrast had the highest numbers of total thrips at 14.6, 15.6 and 16.4/ear, respectively, while Alitis and Anthus had moderate population levels of 5.2

and 6.0/ear, respectively. The lowest number of thrips was found in Robigus, Skater and Marzurka at 0.8, 1.2 and 1.3/ear, respectively (Fig. 22A).

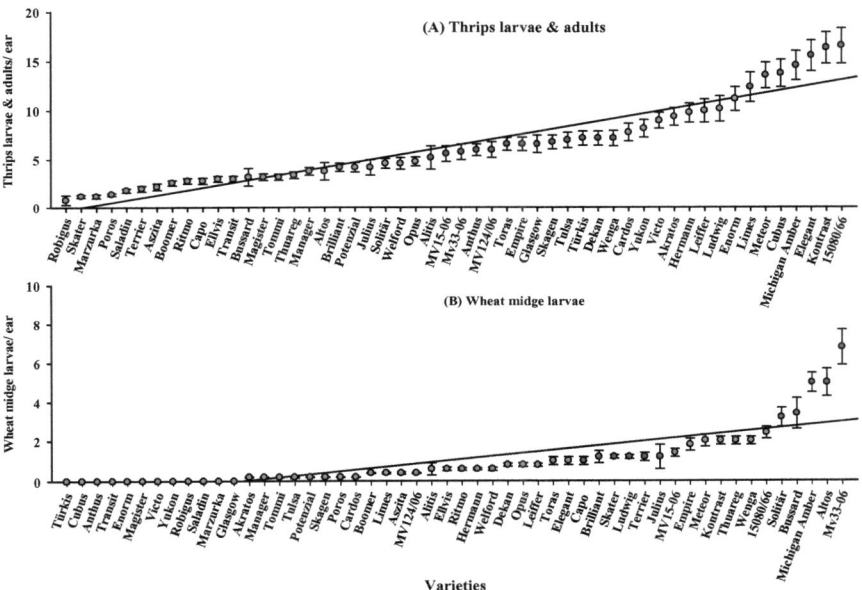

Fig. 22 Mean ± SE of thrips larvae & adults (A) and wheat midges (B) per wheat ear in 52 winter wheat varieties (ordered based on insect infestation) in milky stage (GS 73) in 2009

Wheat midge larvae

There was a significant difference ($P= 0.0475$) in the numbers of both wheat midge larvae per ear among varieties. The highest number of WBM larvae were found in Bussard, Michigan Amber and Altos at 3.4, 5.0 and 5.1 larvae/ear, respectively, while moderate WBM population levels were recorded in Alitis, Ellvis and Ritmo (0.6 larvae/ ear). The variety Akratos had a low population of WBM at 0.2/ear. There were 11 varieties which had no larvae, including Türkis, Cubus, Anthus, Transit, Enorm, Magister, Victo, Marzurka, Yukon, Robigus and Glasgow. The last two are resistant varieties (Fig 22B).

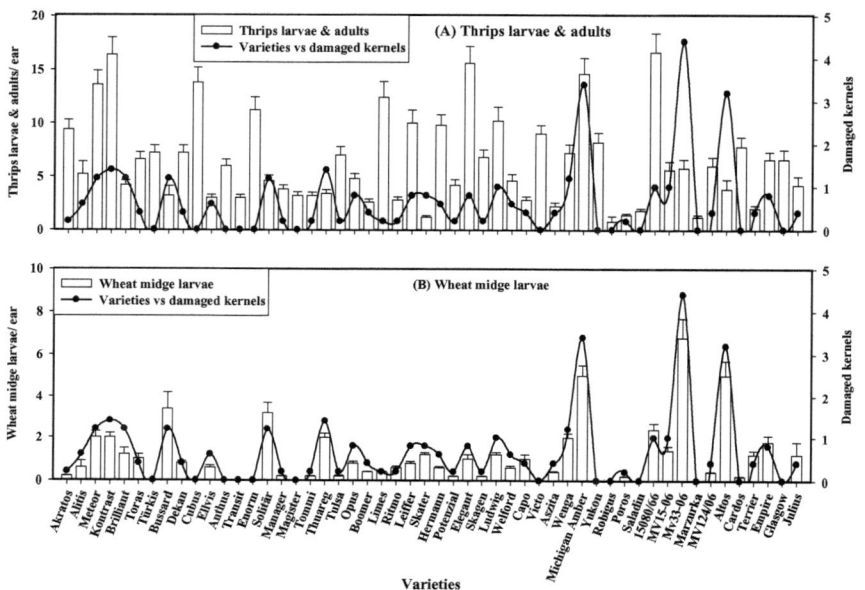

Fig. 23 Mean ± SE of thrips larvae & adults (A) and wheat midges (B) per wheat ear and their correlation to damaged kernels in 52 winter wheat varieties in milky stage (GS 73) in 2009

There were significant differences ($P= 0.0019$) in the number of kernels damaged by thrips and both wheat midges among varieties. Altos and Michigan Amber varieties had the highest number of damaged kernels at 3.2 and 3.4/ear, respectively, while Capo and Aszita had moderate population levels (0.4/ear). The 11 varieties that were wheat midge free, did not have any damaged kernels (Fig 23). There was positive correlation between wheat midge infestation and damaged kernels ($r = +0.76$), while weak significant correlation ($r = +0.17$) was found between thrips and damaged kernels in total thrips (Table 5).

6.2.2.3. Comparison between 2008 and 2009

Generally, the overall density of ear insects (thrips and WBM) was higher in 2008 than 2009. There was a significant difference ($P= 0.0019$) in the number of thrips and WBM per wheat ear between years. The highest number of total thrips were recorded in Ritmo variety

(69.6/ear) in 2008 and Kontrast in 2009 (16.4/ear), while the least number of total thrips were recorded in Thuareg variety (8 larvae/ ear) in 2008 and Robigus (0.8 larvae / ear) in 2009.

There was a significant difference ($P= 0.0021$) in the number of midge larvae per wheat ear between years. The highest WBM numbers were found in Michigan Amber variety (23.8 larvae/ ear) in 2008 and Altos (5.1/ear) in 2009, while the least WBM numbers were observed in Boomer and Akratos varieties (0.2/ear) in both years. WBM larvae were not recorded in both years in the following varieties, Türkis, Anthus, Transit, Enorm, Magister, Victo, and Marzurka. In addition, Welford was WBM free in 2008 and Robigus and Glasgow were WBM free in 2009; these varieties are resistant to wheat midge.

The proportion of ears damaged with wheat midges also differed significantly ($P= 0.0025$) between years. The number of both midge larvae per ear were significantly positively correlated ($r = +0.75$) with the percentage of damaged ears. There was no significant correlation between total thrips and damaged kernels ($r = +0.15$) (Table 5).

6.2.3. Wheat midge larvae population inspected using water traps
6.2.3.1. 2008
S. mosellana

Generally, there were significant differences in *S. mosellana* larvae among varieties and between both years. There was a significant difference ($P= 0.0023$) in the number of orange midge larvae per trap among varieties. Yukon, Tulsa and Saladin varieties had the highest numbers of *S. mosellana* larvae at 35, 36 and 39 larvae/trap, respectively, while the moderate varieties were Poros, Terrier and Brilliant with values of 15, 16 and 17/trap, respectively. The least number of *S. mosellana* were recorded in Bussard and Magister (3/trap). There were traps in some varieties that did not catch any larvae; Enorm, Robigus and Welford; the last two are *S. mosellana* resistant varieties (Fig 24A).

C. tritici

There was a significant difference ($P= 0.0033$) in the number of *C. tritici* larvae per trap among varieties. Leiffer, Empire and Saladin had the highest numbers of *C. tritici* larvae at 11, 12 and 14 larvae/trap, respectively, while moderate *C. tritici* numbers were recorded in Hermann, Alitis and Transit varieties (4/ trap). The least number of *C. tritici* was recorded in Marzurka (1/trap). There were some varieties with no larvae in their traps such as Victo, Capo, Brilliant, Tommi, Boomer, Opus, Dekan, Robigus and Welford (Fig 24A).

Wheat blossom midges caught on different wheat growth stages

Total of wheat midge larvae from all the studied varieties were summed during different growth stages (GS 71-89). Population density was significantly higher ($P= 0.021$) on growth stages 73, 75-77 and 83-85 than others. The population densities of *S. mosellana* at these three growth stages were 24, 444 and 204, respectively; while those for *C. tritici* were 160, 28 and 7, respectively. The last WBM larvae were caught on growth stage 87 (Fig 25A).

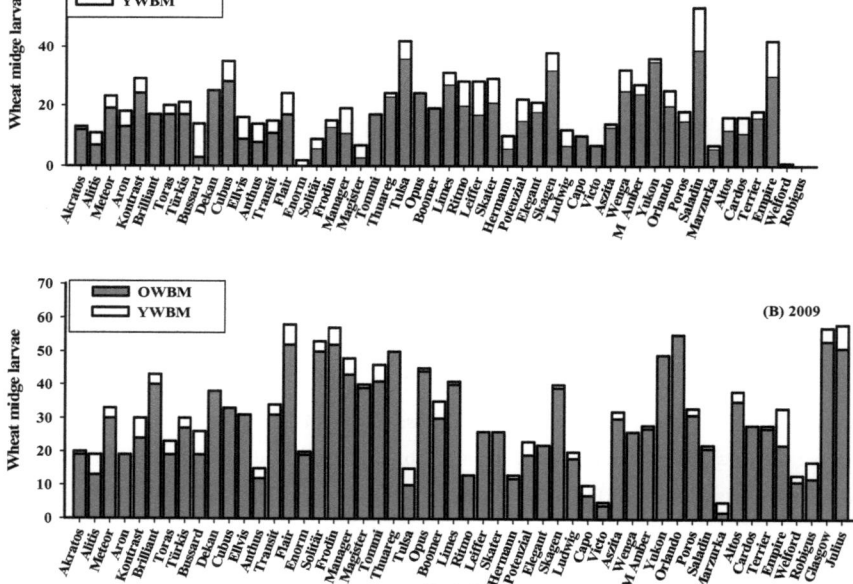

Fig. 24 Total number of *S. mosellana* and *C. tritici* larvae collected in white water traps in selected wheat varieties during 2008 (A) and 2009 (B).

6.2.3.2. 2009

S. mosellana

Larval populations for both wheat midge species were higher in 2009 than in 2008. There was a significant difference ($P= 0.0011$) in the number of *S. mosellana* larvae per trap among varieties. Frodin, Glasgow and Orlando had the highest numbers at 52, 53 and 55 larvae/trap, respectively, while the moderate *S. mosellana* numbers were recorded in Wenga, Türkis and Terrier (27/trap). The least number of *S. mosellana* were recorded in Marzurka and Victo (2 and 4/ trap) (Fig 24B).

C. tritici

There was a significant difference ($P= 0.0021$) in number of *C. tritici* larvae per trap among varieties. Bussard, Julius and Empire had the highest numbers at 7, 7 and 11 larvae/trap, respectively, while the moderate varieties were Ludwig, Aszita and Poros with a value of 2/ trap. The least number of *C. tritici* was recorded in Vitco (1/trap). There were 11 varieties which had no larvae in their traps, including Ritmo, Elegant, Leiffer, Skater, Wenga, Cardos, Ellvis, Cubus, Dekan, Yukon and Thuareg (Fig 24B).

Wheat blossom midges caught on different wheat growth stages

Total of wheat midge larvae were significantly higher ($P= 0.019$) in some growth stages than others. The highest number of *S. mosellana* larvae was recorded at Growth Stages 75-77, 77 and 83, (303, 497 and 336, respectively). The highest number of *C. tritici* was recorded at Growth Stages 71, 73 and 77 (50, 45 and 29, respectively). The last wheat midge larvae were caught on GS 89 (Fig 25B).

6.2.4. Groupings of wheat varieties based on infestation levels

Based on Critical Value for Comparison (Table 6A), the winter wheat varieties could be grouped into three categories for thrip populations (high, moderate and low) and four categories for both wheat midge populations (high, moderate, low and none) (Table 6B).

Table 6A Critical values for comparison between winter wheat varieties in Halle 2008 and 2009

Studied years	Thrips larvae & adults	Wheat midge larvae
2008	25.579*	11.540
2009	12.182	4.1638

*Critical Value for Comparison obtained from using Statistix 9, GLM- ANOVA, and then test of comparison with the best (least infestation) to compare among wheat varieties with the lest infestation variety.

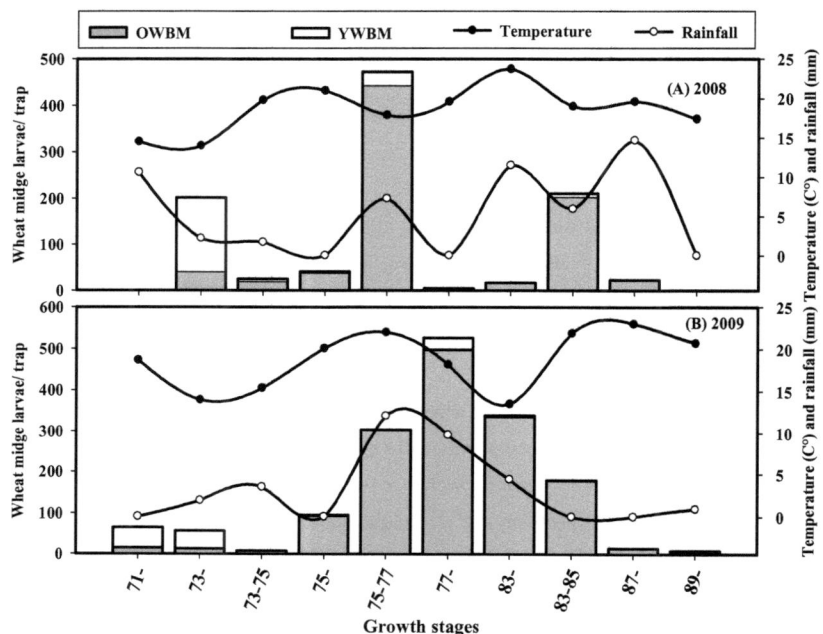

Fig. 25 Total of *S. mosellana* and *C. tritici* larvae collected in water traps and their relation to temperature and rainfall during winter wheat seasons 2008 (A) and 2009 (B).

Table 6B Groupings of wheat varieties based on infestation levels of wheat ear insects in Halle

Infestation grade	Thrips	Wheat midges
High	Akratos, Limes, Ritmo, Michigan Amber, Elegant, Meteor, Kontrast, Cardos, Opus, Yukon and Manager.	Michigan Amber, Altos, Skater, Terrier, Tulsa, Frodin, Orlando, Julius, Leiffer, and Saladin
Moderate	Enorm, Cubus, Tommi, Alitis and Anthus, Transit, Saladin, Bussard, Hermann, Capo, Türkis, Welford, Julius, Solitär, Altos, Toras, Tulsa, Victo, Wenga and Empire	Alitis, Ellvis, Ritmo, Potenzial, Skagen, Bussard, Poros, Brilliant, Wenga, Ludwig, Aszita, Poros Hermann, Empire, Tommi, and Dekan.
Low	Thuareg, Robigus, Skater, Marzurka, Poros, Terrier, Aszita, Boomer, Brilliant, Leiffer, Ludwig, Ellvis, Dekan, Potenzial and Skagen	Akratos, Boomer, Türkis, Cubus, Anthus, Transit, Yukon, Solitär, Elegant, Cardos, Thuareg, Opus and Glasgow
None	-	Capo, Vitco, Enorm, Magister, Marzurka Welford, Robigus

6.3. COMPARISON BETWEEN EAR INSECT INFESTATION LEVELS 2008

According to ear insect infestation in Halle and Silstedt, this infestation was higher in 2008 than 2009; therefore the comparison between two sites was done only in 2008.

6.3.1. Halle site

There was significant difference ($P= 0.046$) in the number of thrips adults per ear among varieties. Tommi and Welford varieties had the highest numbers of thrips adults 7.6 and 8.2/ ear, respectively, then followed by Türkis (6 individuals/ ear), Anthus, Robigus, Potenzial and Boomer (4.8, 4.8 and 3.4 individuals/ ear), lastly, Dekan (1.8 individuals/ ear) (Fig. 26).

There was significant difference ($P= 0.049$) in the number of thrips larvae per ear among varieties. Türkis and Welford varieties had the highest numbers of thrips larvae 28.0 and 26.4/ ear, respectively, then followed by Tommi (20.6 individuals/ ear), Anthus, Robigus, Dekan and Boomer (18.2, 18.6, 18.6 and 17.4 individuals/ ear), lastly, Potenzial (12.6 individuals/ ear) (Fig. 26).

There was significant difference ($P= 0.027$) in the number of total thrips per ear among varieties. Türkis and Welford varieties had the highest numbers of total thrips 32.0 and 34.6/ ear, respectively, then followed by Tommi (28.4 individuals/ ear), Anthus, Robigus, Dekan and Boomer (23.0, 23.4, 20.4 and 20.8 individuals/ ear), lastly, Potenzial (17.4 individuals/ ear) (Fig. 26).

There was significant difference ($P= 0.045$) in the number of wheat midges larvae per ear among varieties. Tommi and Potenzial varieties had the highest numbers of WBM larvae 1.4 and 1.2/ ear, respectively, then followed by Dekan (0.8 individual/ ear), Robigus, and Boomer (0.2/ ear) lastly, Türkis, Anthus and Welford (0 / ear) (Fig 26).

There was significant difference ($P= 0.032$) in the damaged kernels resulted from thrips or wheat midges among varieties. Tommi and Potenzial varieties had the highest damaged kernels 0.4 and 1.0/ ear, respectively, then followed by Dekan, Robigus, and Boomer (0.2 individual/ ear) lastly, Türkis, Anthus and Welford (0/ ear) (Fig. 26). There is a correlation between wheat midge infestation and damaged kernels ($r= +0.79$), while no correlation between thrips and damaged kernels ($r= +0.02$, -0.26 and -0.22) in thrip adults, larvae and total thrips, respectively (Table 7).

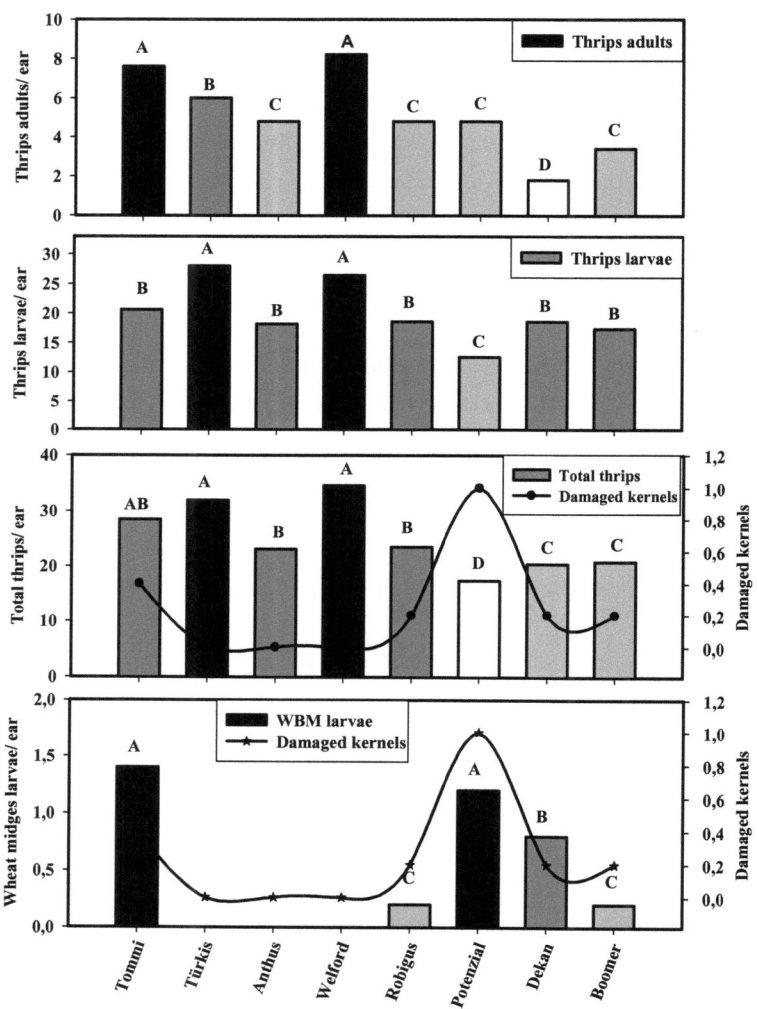

Fig. 26 Mean of thrips adults, larvae and total thrips, wheat midge larvae and the relation to damaged kernels in different winter wheat cultivars (Growth stage 73) in Halle 2008
Different letters and colors indicate significant differences

6.3.2. Silstedt site

There was significant difference ($P= 0.009$) in the number of thrips adults per ear among varieties. Türkis and Anthus varieties had the highest numbers of thrips adults 5.5 and 5.1/ ear, respectively, then followed by Boomer, Tommi and Dekan (3.9, 2.6 and 2.4 individuals/ ear), lastly, Welford, Robigus and Potenzial (2.0, 1.9 and 1.8 individuals/ ear) (Fig. 27).

There was significant difference ($P= 0.0019$) in the number of thrips larvae per ear among varieties. Türkis and Anthus varieties had the highest numbers of thrips larvae 12.2 and 12.1/ ear, respectively, then followed by Tommi and Boomer (6.7 and 6.4 individuals/ ear), Dekan, Welford and Potenzial (4.6, 4.0 and 3.5 individuals/ ear), lastly, Robigus (1.3 individuals/ ear) (Fig. 27).

There was significant difference ($P= 0.0003$) in the number of total thrips per ear among varieties. Türkis and Anthus varieties had the highest numbers of total thrips 17.7 and 17.2/ ear, respectively, then followed by Tommi and Boomer (11.1 and 10.3 individuals/ ear), Dekan, Welford and Potenzial (7.0, 6.0 and 5.3/ ear), lastly, Robigus (3.2 individuals/ ear) (Fig. 27).

There was significant difference ($P= 0.002$) in the number of wheat midges larvae per ear among varieties. Türkis and Dekan varieties had the highest numbers of WBM larvae 15.5 and 16.0/ ear, respectively, then followed by Tommi, Potenzial and Boomer (12.4, 10.7 and 10.4 individuals/ ear), Anthus (4.3 individuals/ ear), lastly, Welford and Robigus (2.3& 1.5 individuals/ ear). The later are 2 resistant varieties (Fig. 27).

There was significant difference ($P= 0.001$) in the damaged kernels resulted from thrips or wheat midges among varieties. Türkis, Tommi and Dekan varieties had the highest damaged kernels 8.6, 7.8 and 7.8/ ear, respectively, then followed by Potenzial and Boomer (6.6 and 6.7 individuals/ ear), Anthus (3.2 individuals/ ear) lastly, Welford and Robigus (0.9 and 1.0 individuals/ ear) (Fig. 27). There is a correlation between WBM infestation and damaged kernels ($r= +0.94$), while no correlation between thrips and damaged kernels ($r= +0.12, +0.17$ and $+0.16$) in thrip adults, larvae and total thrips, respectively (Table 7).

Table 7 Correlation coefficient between ear insects (wheat midges& thrips) and damaged kernels in Halle and Silstedt 2008

Sites	Thrips adults	Thrips larvae	Total thrips	Wheat midges
Halle	+0.02	-0.26	-0.22	+0.79 *
Silstedt	+0.12	+0.17	+0.16	+0.94 *

* Significant differences are at 0.05 levels

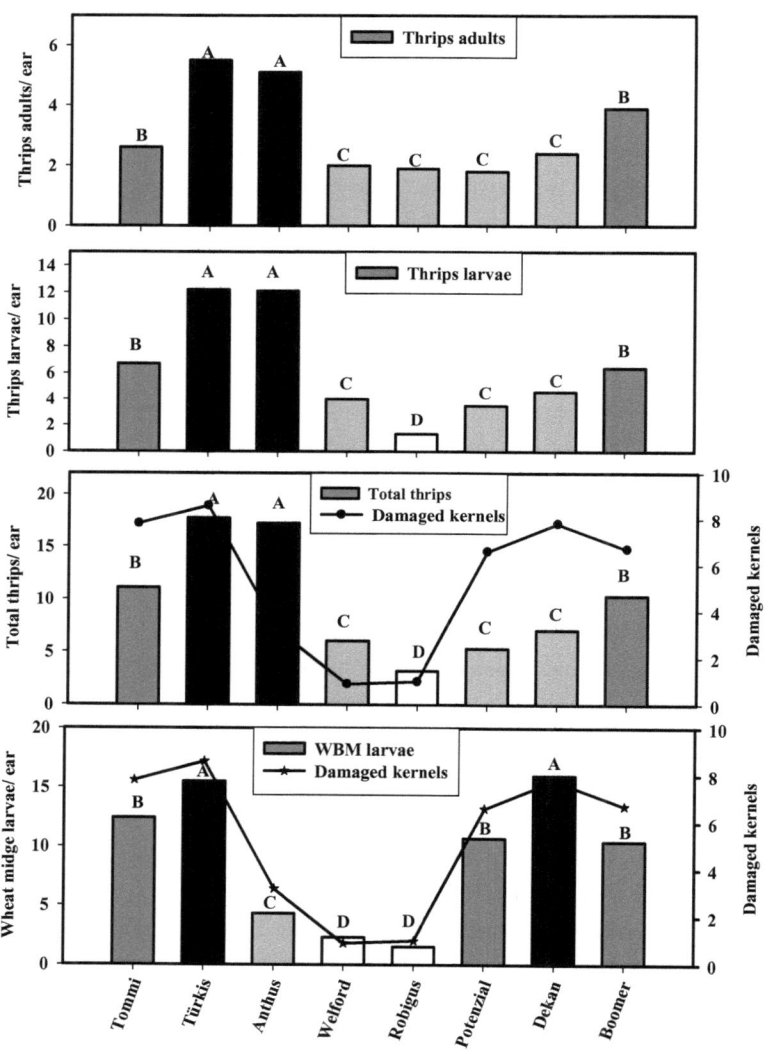

Fig. 27 Mean of thrips adults, larvae and total thrips, wheat midge larvae and the relation to damaged kernels in different winter wheat cultivars (Growth stage 73) in Silstedt 2008
Different letters and colors indicated significant differences

6.3.3. Comparison between Halle and Silstedt

There were significant differences in thrips and WBM among varieties in both sites Halle and Silstedt. Numbers of thrips adults were higher significantly ($P= 0.001$) in varieties Tommi and Welford in Halle and Türkis and Anthus in Silstedt, while the least thrips' adults numbers were recorded in varieties Dekan and Boomer in Halle and Robigus and Potenzial in Silstedt. Thrips larvae and total thrips were significantly higher ($P= 0.001$) in varieties Türkis and Welford in Halle and Türkis and Anthus in Silstedt. The least thrips populations were recorded in varieties Potenzial and Boomer in Halle and Robigus and Potenzial in Silstedt (Fig. 28).

There were significant differences in the midge larvae. Their numbers were higher significantly ($P= 0.001$) in varieties Tommi and Potenzial in Halle and Türkis and Dekan in Silstedt, while the least WBM numbers were observed in varieties Anthus, Welford and Robigus in Halle and Silstedt. Welford and Robigus varieties are resistant varieties in both sites (Fig. 28).

The proportion of ears damaged with wheat midges also differed significantly ($P= 0.001$) among varieties and between both fields. The number of midge larvae per ear was significantly positively correlated ($P= 0.005$, $r= +0.96$) with the percentage of ears infested. There was no significant correlation between the number of thrips and damaged kernels (Fig. 28).

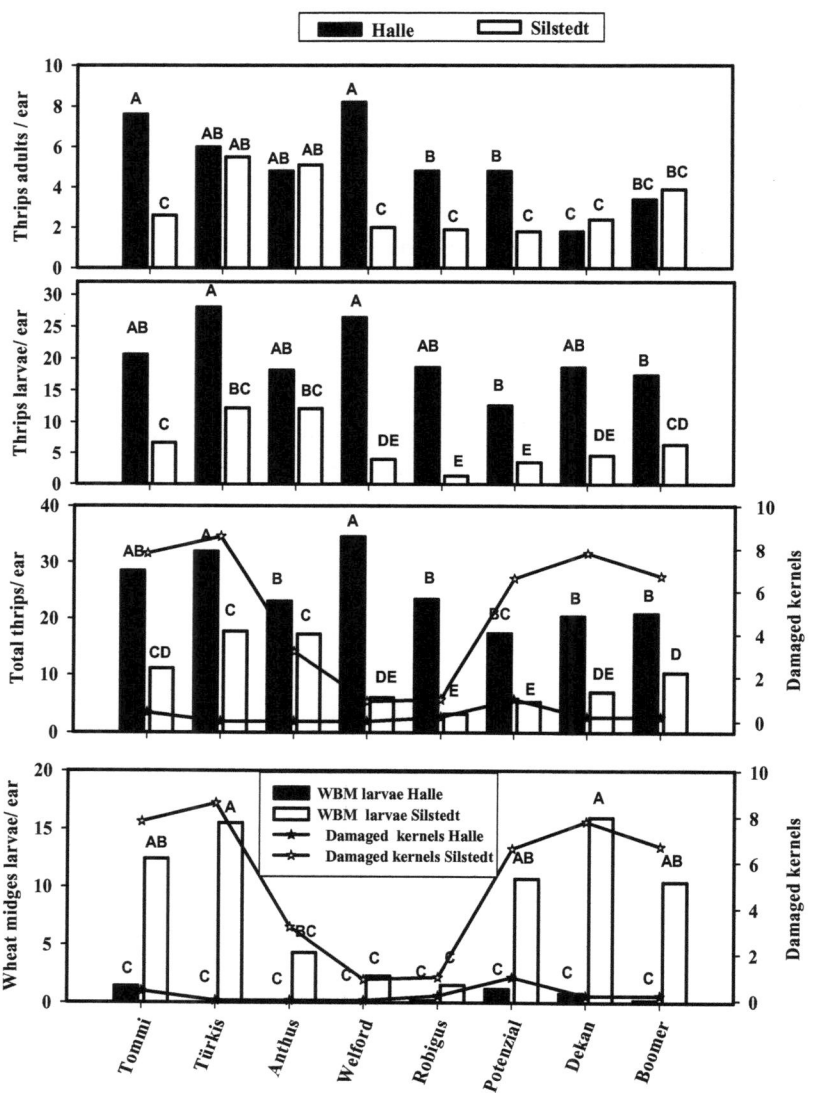

Fig. 28 Comparison between Halle and Silstedt in mean of thrips adults, larvae and total thrips, WBM larvae and the relation to damaged kernels in different winter wheat cultivars (Growth stage 73) in 2008
Different letters indicated significant differences

6.4. SPRING WHEAT VARIETIES IN HALLE
6.4.1. Population of *S. mosellana* adult surveyed using pheromone traps

Orange wheat midge population started slowly in 2008. There was one peak which recorded on GS 47-51 (128 males/ trap). The population went slowly down till taking traps off at Gs 77 (Fig. 29A). In 2009, *S. mosellana* male population started early; one peak was recorded (18 midges/trap) at GS 51-55 (Fig. 29B). Orange wheat midge population was higher in 2008 than 2009. Coincidence between adult flight and susceptible growth stages 49-65 allowed a higher infestation to develop from low population levels. There was a strong correlation between the peak of pheromone trap catches and temperature and rainfall ($r = 0.87$ and $r = 0.71$) in 2008 and 2009, respectively, because dry weather (low rainfall >2mm/day and high temperature >17°C) as recorded at GS 47 to 51 help midge adults to hatch from cocoon in overwintering, therefore the highest populations were recorded at GS 47-51 in 2008 (Fig. 29A) and at GS 51-55 in 2009 (Fig. 29B). Coincidence of adult activity and susceptible growth stages was more obvious in 2008 than in 2009 as shown in oval shapes in Fig 29A&B.

6.4.2. Evaluation of thrips and midges in wheat ears
6.4.2.1. 2008
Total thrips (larvae and adults)

In flowering stage (GS 65), there was no significant difference ($P= 0.145$) between Sakha 93 and Triso in thrips density. In milky stage (GS 73), significant differences were found ($P= 0.011$) in the number of thrips between both varieties. Triso variety had higher numbers of thrips (19.5 / ear) than Sakha 93 (5.8 individuals / ear) (Fig. 30A).

Wheat midges

Wheat midge larvae were found in the flowering stage with small numbers and there was no significant difference ($P= 0.055$) between the varieties (Fig 30B). In milky stage (GS 73) a significant difference ($P= 0.002$) was observed in the number of wheat midge larvae per ear between Sakha 93 with (0.7/ear) and Triso (12.3/ear) (Fig. 30B).

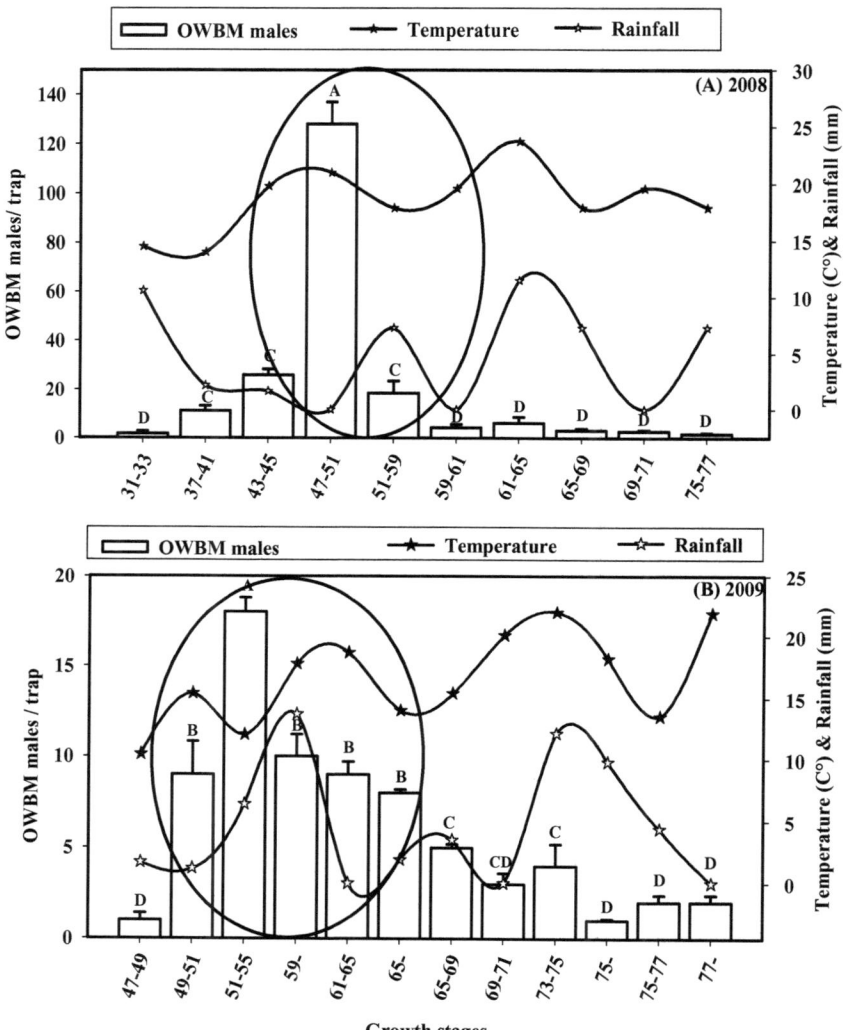

Fig. 29 Mean ± SE of *S. mosellana* male catches in pheromone traps and their relation to temperature and rainfall in spring wheat 2008 (A) & 2009 (B). Ovals refer to coincidence of adult activity and susceptible growth stages. Different letters indicate significant differences.

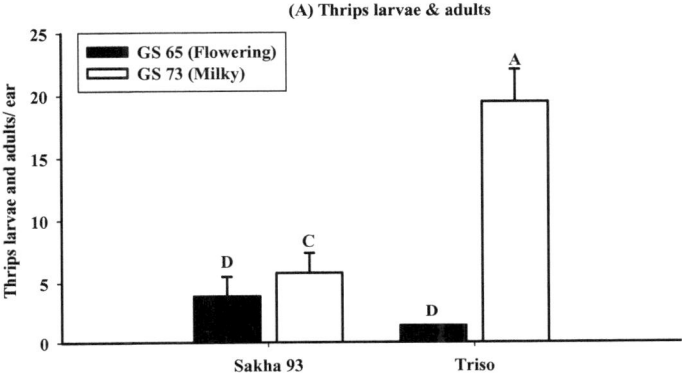

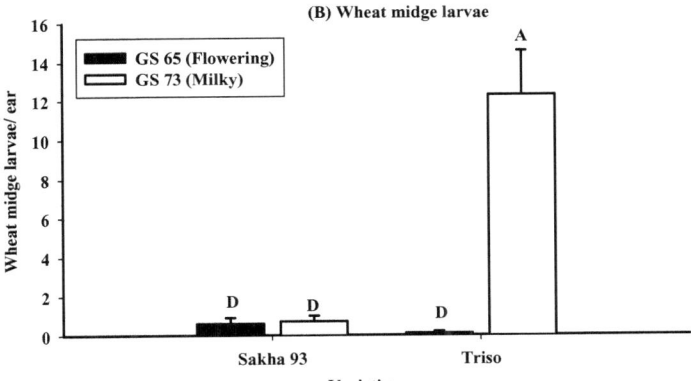

Fig. 30 Mean ± SE of thrips larvae & adults (A) and wheat midge larvae (B) in two growth stages 65 and 73 in spring wheat varieties (Sakha 93 and Triso) in 2008. Different letters indicate significant differences

6.4.2.2. 2009

Thrips (larvae and adults)

Thrips density was higher in 2009 than 2008. In flowering stage (GS 65), there was no significant difference ($P= 0.280$) between Sakha 93 and Triso in thrips numbers. In milky stage (GS 73) significant differences were recorded ($P= 0.034$) in the number of thrips between both varieties. Triso variety had higher thrips numbers (32.1/ ear) than Sakha 93 (23.1 individuals/ ear) (Fig. 31A).

Wheat midge larvae

Wheat midges were lower in 2009 than in 2008. Small numbers of midge larvae were recorded at the flowering stage with and was no significant different ($P= 0.534$) between the two varieties. There was a significant difference ($P= 0.025$) in the number of wheat midge larvae per ear between Sakha 93 and Triso, in milky stage (GS 73); Sakha 93 had lower midge larvae (3.6/ ear) content than Triso (6.5 individuals/ ear) (Fig. 31B).

6.4.2.3 Correlation between numbers of thrips, midges and damaged kernels

There were significant differences ($P= 0.001$ & $P= 0.045$) in damaged kernels in both years, respectively, resulting from thrips and wheat midge infestation. Sakha 93 had lower damaged kernels than Triso variety. In 2008, the wheat midge larvae numbers and damaged kernels ($r= +0.88$ and $+0.69$) were correlated in GS 65 and 73 stages respectively; while there was no correlation between total thrips and damaged kernels for both stages. In 2009, the correlation value between the number of midge larvae and damaged kernel was $r= +0.99$ in both growth stages; while no significant correlations between total thrips and damaged kernels in both stages was found (Table 8).

Table 8 Correlation coefficients between number of ear insects (thrips and wheat midges) and number of damaged kernels in flowering and milky stages in Halle 2008 and 2009

Years	Growth stages	Thrips larvae & adults	Wheat midge larvae
2008	Flowering stage (GS 65)	+0.20	+0.88 *
	Milky stage (GS 73)	+0.21	+0.69*
2009	Flowering stage (GS 65)	+0.23	+0.99*
	Milky stage (GS 73)	+0.27	+0.99*

* Significant differences are at 0.05 levels

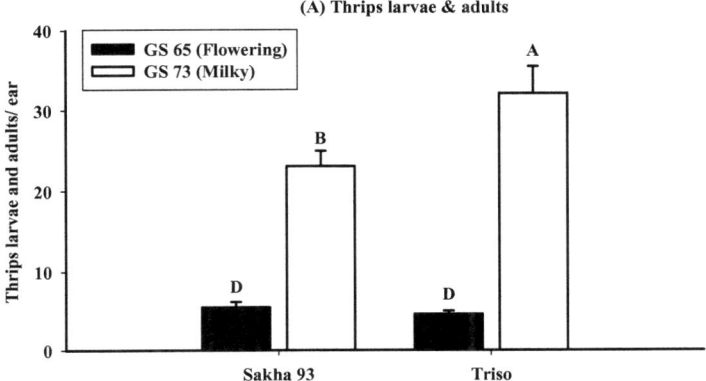

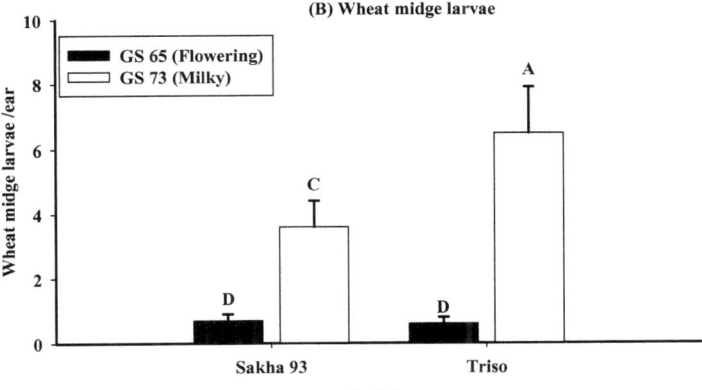

Fig. 31 Mean ± SE of thrips larvae & adults (A) and wheat midge larvae (B) in two growth stages 65 and 73 in spring wheat varieties (Sakha 93 and Triso) in 2009. Different letters indicate significant differences.

6.4.3. Wheat midge larvae population inspected using water traps

6.4.3.1. 2008

Orange wheat midge (*S. mosellana*)

S. mosellana population sampled was significantly higher ($P = 0.008$) on the two sampling dates of 26th June and 7th July (GS 75 and 77-79) than on other dates. The last larvae were caught on 23rd July (GS 87) (Fig. 32A). Analyses of the cumulative data (ANOVA) to compare total *S. mosellana* larvae numbers showed that there was no significant difference ($P= 0.259$) between the two varieties but generally the total of caught larvae numbers were higher in Triso than in the variety Sakha 93 (Fig. 34A).

Yellow wheat midge (*C. tritici*)

The population was significantly higher ($P= 0.001$) on the two sampling dates 26th June and 7th July (GS 75 and 77-79) than on other dates. The last *C. tritici* larvae were caught on 14th July in Sakha 93 (GS 83) and on 23rd July in Triso (GS 87) (Fig. 32B). Analyses of the cumulative data using ANOVA to compare total *C. tritici* larvae numbers showed significant difference ($P= 0.004$) between Sakha 93 and Triso (Fig. 34A).

6.4.3.2. 2009

Orange wheat midge (*S. mosellana*)

Populations of *S. mosellana* larvae were higher in 2009 than 2008. Population density was significantly higher ($P= 0.003$) on the three sampling dates 7th, 10th and 14th July than on other dates. The results indicated, that emigrated *S. mosellana* larvae could be divided into three groups; the high populations were recorded on 7th, 10th and 14th July (GS 75, 77 and 79), followed by middle populations which recorded on 16th, 21st and 27th July (GS 81, 83 and 85) and the low populations were recorded in the early and end of season (Fig. 33A). The last *S. mosellana* larvae were caught on 3rd August (GS 89) (Fig. 33A). Analyses of the cumulative data using ANOVA showed that there was significant difference ($P= 0.001$) between the two varieties (Fig. 34B)

Yellow wheat midge (*C. tritici*)

The catches of *C. tritici* larvae were higher in 2008 than 2009. Population density was significantly higher ($P= 0.002$) on the three sampling dates 7th, 10th and 14th July than on the other dates. The results indicated that *C. tritici* larvae emigration could be divided into two groups; high populations were recorded on 7th, 10th and 14th July (GS 75, 77 and 79) and the low populations were recorded on 16th, 21st and 27th July (GS 81, 83 and 85), while no *C. tritici* was recorded in the early and end of season (Fig. 33B). The last *C. tritici* larvae were caught on 30th July in Sakha 93 (GS 87) and on 27th July in Triso (GS 85) (Fig. 33B). ANOVA analyses of the cumulative data using the total *C. tritici* larvae numbers showed that there was a significant difference ($P= 0.009$) between Sakha 93 and Triso (Fig. 34B).

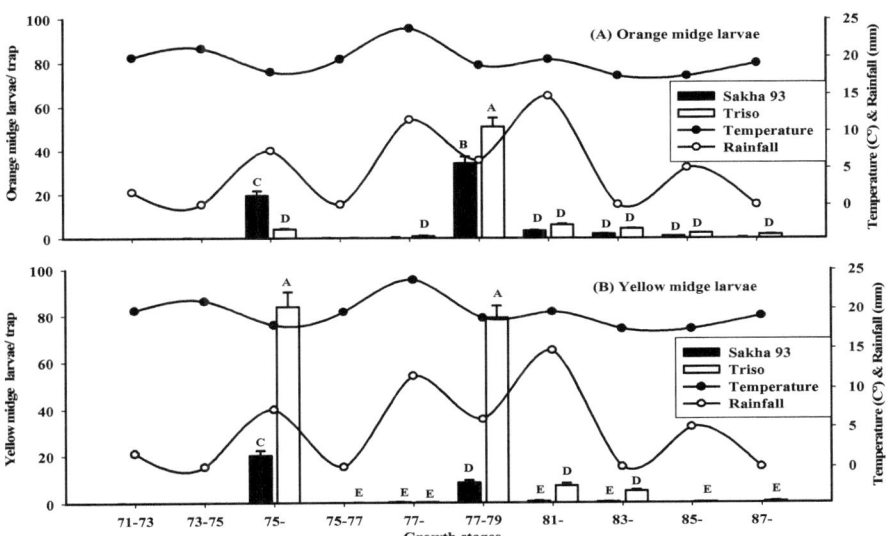

Fig.32 Mean ± SE of orange (A) and yellow (B) wheat midge larvae catches/ water trap and their relation to temperature and rainfall in two spring wheat varieties during different growth stages 2008. Different letters indicate significant differences.

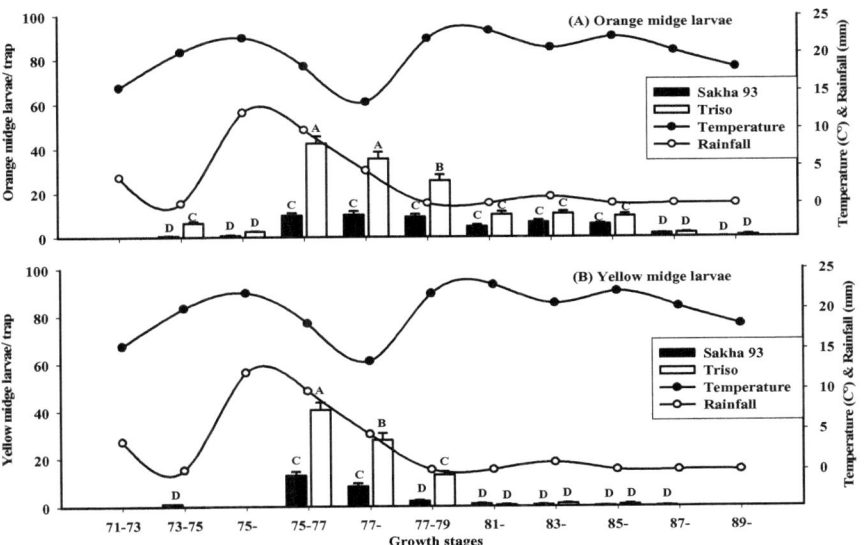

Fig. 33 Mean ± SE of orange (A) and yellow (B) wheat midge larvae catches/ water trap and their relation to temperature and rainfall in two spring wheat varieties during different growth stages 2009. Different letters indicate significant differences.

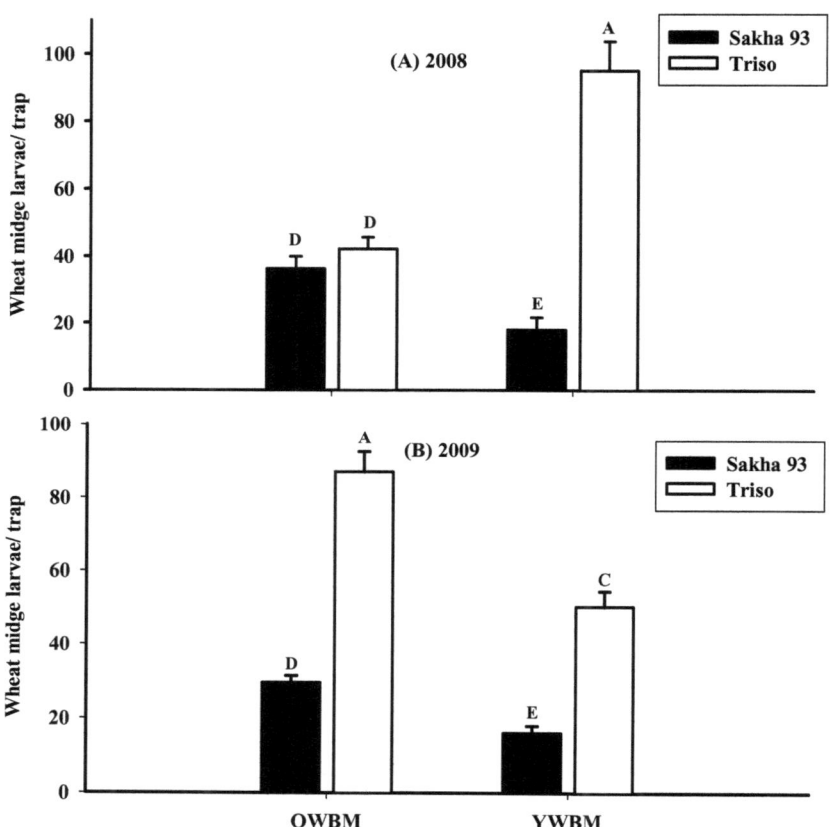

Fig. 34 Mean ± SE of orange and yellow wheat midge larvae catches/ water trap in two spring varieties in 2008 (A) and in 2009 (B). Different letters indicate significant differences.

6.4.4. Spring wheat yield

6.4.4.1. 2008

The analysis of yield data implies a significant difference ($P= 0.0033$) between the two varieties; comparison of yields indicated that the Triso outyielded Sakha 93 variety, except weight of 1000 kernels; it was heavier in Sakha 93 than Triso. The yield indices in the Triso variety was an average of 49.3 kernels/ear, 1.98 kg kernels/plot and 4416 kg/ha, while in the

Sakha 93 variety was 41.3 kernels/ear and, 1.6 kg/ plot and 3564 kg/ ha. The weight of 1000 kernels, was for Triso 45.1g and for Sakha 93 51.2g respectively (Table 9).

6.4.4.2. 2009

The analysis of yield data showed 2009 again significant differences ($P= 0.0021$) between two varieties. The yield indices in the Triso variety was in the mean 49.5 kernels/ ear, 43.8g/ear and 2.02 kg/ plot and 4483 kg/ ha, while in the Sakha 93 variety the means were 40.5 kernels/ ear, 52.9g/ ear, 1.5 kg/ plot and 3395 kg/ ha respectively (Table 9).

Table 9 Mean ± SE of yield index in spring wheat varieties (kernel numbers/ear, weight of 1000 kernels, weight of grains/ plot and weight of grains/ ha) in Halle 2008 and 2009

Years	Varieties	Kernel numbers/ ear	Weight of 1000 kernels (g)	Weight of grain/ plot (Kg)	Weight of grains/ ha (Kg)
2008	Triso	49.32 ± 4.9 A	45.14 ± 1.2 B	1.98 ± 0.2 A	4416 ± 413 A
	Sakha 93	41.32 ± 1.1 B	51.25 ± 3.1 A	1.60 ± 0.1 B	3564 ± 247 B
2009	Triso	49.54 ± 3.6 A	43.80 ± 1.5 B	2.02 ± 0.1 A	4483 ± 524 A
	Sakha 93	40.52 ± 1.9 B	52.91 ± 2.9 A	1.53 ± 0.2 B	3395 ± 525 B

* Different letters indicate significant differences.

6.4.5. Practical recommendation: pheromone traps, ear insect evaluation and water traps

Pheromone catches were significantly different ($P= 0.0243$) between Sakha 93 and Triso. Mean of these catches during the susceptible period (GS 49-65) were 50 & 65 in 2008 and 7 & 11 adults/ trap in 2009 in both wheat varieties, respectively.

We could conclude that threshold of *S. mosellana* in susceptible stages were >30 adults/ pheromone trap/ day in case of stable whether of temperature (> 17C°), light wind and no rainfall. From both results of ear insects evaluation and water traps catches, the relation of the number of wheat midge larvae which evaluated in 10 wheat ears to the number of larvae found in the water traps were 24% and 90% in 2008 and 52% and 94% in 2009. These percents of wandered larvae to soil and surveyed by water traps (circle surface area of each trap= 122.5 cm^2) in the respective growth stage (milky stage GS 71-79) as shown in Table (10). We could also conclude that threshold of *S. mosellana* plus *C. tritici* in migration stage were >40 larvae/ water trap/ day after heavy rain. This is considered a serious prognosis for crop rotation in the next year in case of intending cultivation wheat after wheat. Monitoring technique is necessary to use in the beginning of season in the study region (Central Germany). Wheat midges can cause damage to spring wheat that reaches economic levels in central Germany, as well as in winter wheat. Overall the results suggest that a trapping program based on pheromone and water traps should be the best way of sampling wheat midges in the early and end of season.

Table 10 Number of *S. mosellana* adults in pheromone traps, number of wheat midge larvae (red-orange and yellow) in the ear evaluation and in the water traps in two spring wheat varieties in Halle 2008 and 2009.

Years / wheat varieties	2008		2009	
	Sakha 93	**Triso**	**Sakha 93**	**Triso**
$\bar{x}$ Adults / pheromone trap	50± 1.9B	65±2.3 A	7.0±1.0 D	11.0± 1.2 C*
$\sum$ Adults in pheromone traps	185 B	219 A	42.0 D	66.0 C
$\bar{x}$ WBM larvae/ 1 ear	1.30±0.3 D	12.40±1.20 A	4.30±0.5 C	7.10 ± 0.8 B
$\bar{x}$ WBM larvae/ 10 ears	13.0±1.5 D	124.0±6.00 A	43.0±2.6 C	71.0 ± 4.2 B
$\bar{x}$ WBM larvae/ water trap	54.4±7.3 C	137.2±12.2 A	45.6±3.7 C	137.2±9.7 A
% WBM L/ ear: WBM L/ water trap	23.9 C	90.4 A	94.3 A	51.7 B

* Different letters indicate significant differences

6.5. WINTER WHEAT FIELDS IN SALZMÜNDE

6.5.1. Population of *S. mosellana* adult surveyed using pheromone traps

Populations of *S. mosellana* adults started slowly till milky stage and the first peak was recorded at GS 73 (1496 midges/ trap) in 2007 (Fig. 35). In 2008, large variations in numbers of midges in the pheromone traps and in time of peak catches were found; the highest number of males was 173 midges/ trap recorded in GS (55-59) (Fig. 36). There was one peak in 2009 (32 midges/ trap) at GS 59-61 (Fig. 37).

The lowest number of midges were 1, 13 and 2.5 midges/trap in 2007, 2008, and 2009, respectively (Fig. 35, 36, 37). Coincidence of adult activity and susceptible growth stages was more obvious in 2008 than in 2009 as shown in the oval shape in Fig 36 & 37. The susceptible stages of wheat coincided with suitability for flight and oviposition. There was also a strong correlation between peak pheromone trap catches and weather conditions, rainfall and temperature ($r= +0.892$ and $r= +0.742$) in 2008 & 2009, respectively. On the other hand, there was no correlation ($r= +0.38$) in 2007, possibly because the midge activity started later than the susceptible stage.

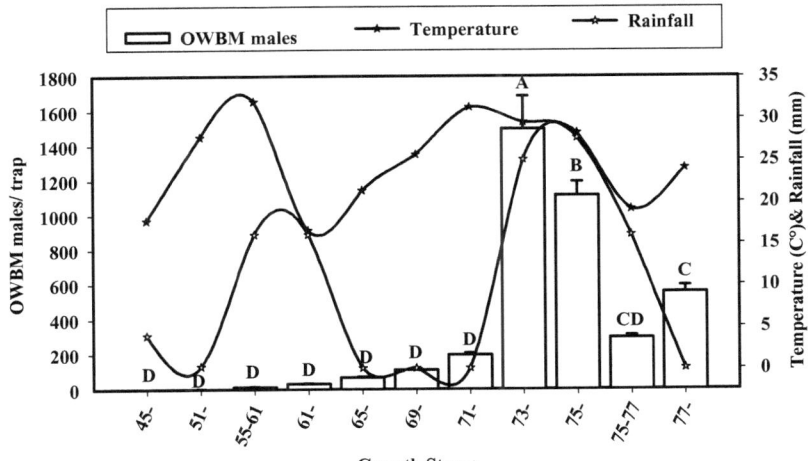

Fig. 35 Mean *S. mosellana* male catches in pheromone traps and their realtion to temperature and rainfall in Salzmünde 2007.Different letters indicate significant differences.

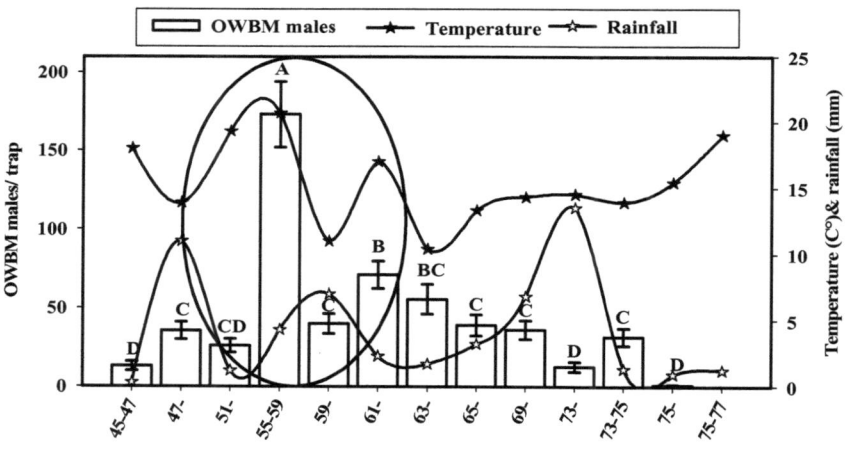

Fig. 36 Mean ± SE of *S. mosellana* male catches in pheromone traps and their relation with temperature and rainfall in 2008. Oval refers to coincidence of adult activity and susceptible growth stages. Different letters indicate significant differences.

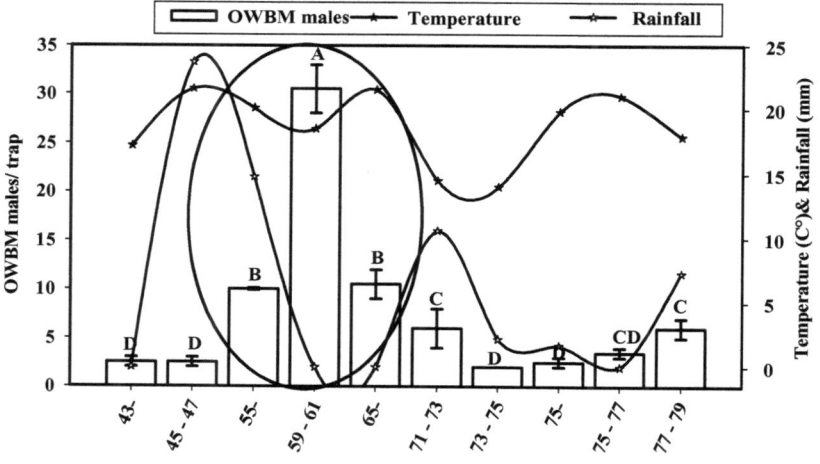

Fig. 37 Mean ± SE of *S. mosellana* male catches in pheromone traps and their relation to temperature and rainfall in 2009. Oval refers to coincidence of adult activity and susceptible growth stages. Different letters indicate significant differences.

6.5.2. Evaluation of thrips and midges in wheat ears

6.5.2.1. 2007

Total thrips (larvae and adults)

In the most important growth stage GS 65 and 73 there was significant difference in thrips populations ($P= 0.0047$) ($P= 0.0484$) and ($P= 0.0451$) in thrips adults, larvae and total thrips, respectively. The thrips adults were 0.7 and 1.5/ ear in the same way. The corresponding records in thrips larvae were 1.3 and 2.0 / ear. The total thrips/ ear were 2.1 and 3.5, respectively (Fig. 38).

Wheat midges

There was a significant difference ($P= 0.0357$) in total midges between both growth stages (flowering and milky). Total midges (*S. mosellana* & *C. tritici*) were 0.2 and 1.8 larvae/ ear, respectively (Fig. 38).

Damaged kernels by thrips and midges

There was significant difference ($P= 0.0391$) in damaged kernels (deformated, cherviled or cracked kernels) between the growth stages 65 and 73, these values were 0.2 and 1.8 damaged kernels/ ear, respectively (Fig. 38).

6.5.2.2. 2008

Total thrips (larvae and adults)

Thrips population was 10.0 thrips/ ear before the insecticide application, while after 3 days post treatment; they were 8.8 and 26.0 thrips/ ear in the treated and control, respectively.

In flowering stage (GS 65)

Significant differences were found ($P= 0.0083$) in the number of total thrips between treated and control. On the 7th day, thrips number in control plants were higher than in treated 26.4 and 6.8 individuals/ ear, respectively; the corresponding numbers on the 10th day were 27.6 and 6.8 thrips/ ear (Fig. 39A).

In milky stage (GS 73)

There was a significant difference ($P= 0.0041$) in thrips number between treated and untreated. Thrips numbers were lower in the treated than control. They were 6.8 and 31.6 thrips/ ear, respectively after 15 days post treatment; the corresponding records on 20th day were 18.4 and 34.4 thrips/ ear (Fig. 39A).

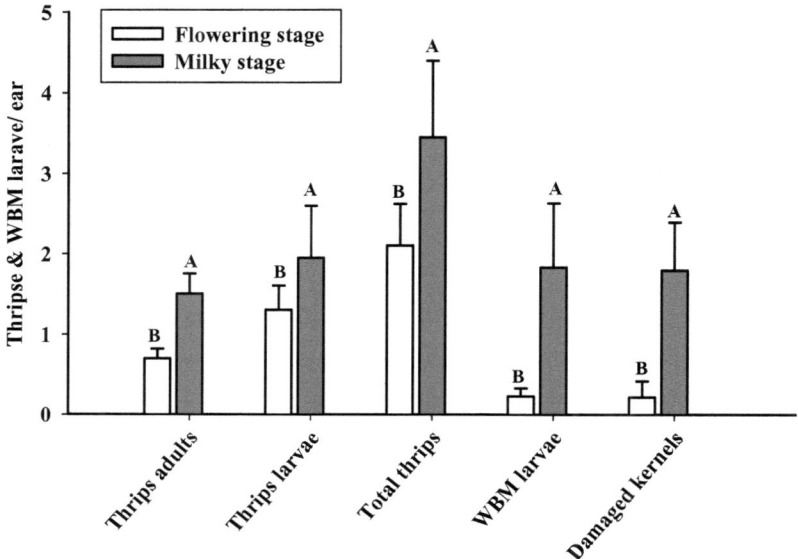

Fig. 38 Mean ± SE of thrips (adults, larvae & total) and WBM midges in two growth stages in 2007. Different letters indicate significant differences.

Wheat midge larvae

There was no wheat midge larvae recorded before treatment (Fig. 39B), while 3 days after treatment; they were 0.0 and 4.4 midge larvae/ ear in treated and control plots, respectively.

In flowering stage (GS 65)

There was no significant difference ($P= 0.0672$) in the number of midge larvae (*S. mosellana* & *C. tritici*) between treated and untreated plots. On the 7th day, midge larvae numbers in treated were lower than in control 0.8 and 2.0 individuals/ ear, correspondingly; the equivalent records on the 10th day were 2.4 and 3.6 thrips/ ear (Fig. 39B).

In milky stage (GS 73)

There was a significant difference ($P= 0.0245$) in wheat midge larvae between treated and untreated plants. Midge larvae number was higher in control than in treated plants. They were 4.0 and 1.2 larvae/ ear, respectively after 15 days post treatment; the corresponding numbers on 20th day were 4.8 and 2.4 larvae/ ear; this meant that treated plants had an half the population recorded on control plants (Fig. 39B).

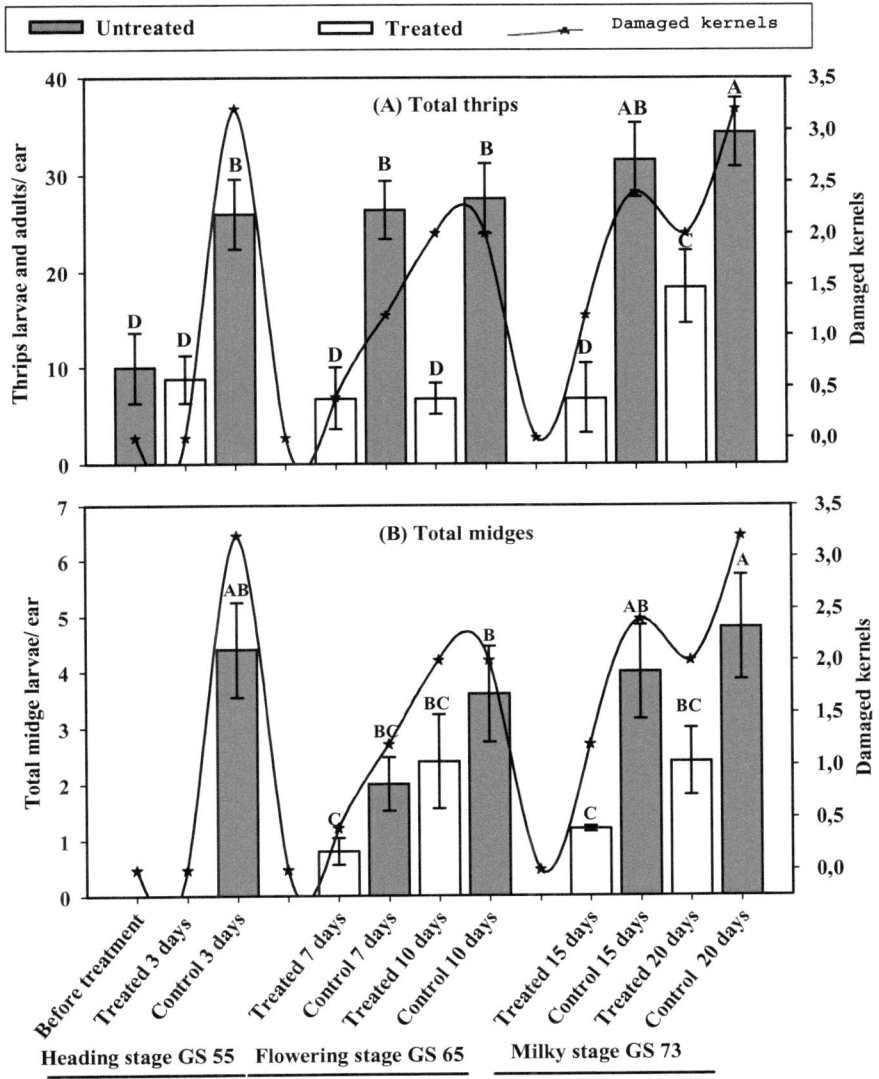

Fig. 39 Mean ± SE of total thrips (A) and midge larvae (B) in treated and untreated winter wheat in 2008 in 3 different growth stages and their correlation to damaged kernels. Different letters indicate significant differences.

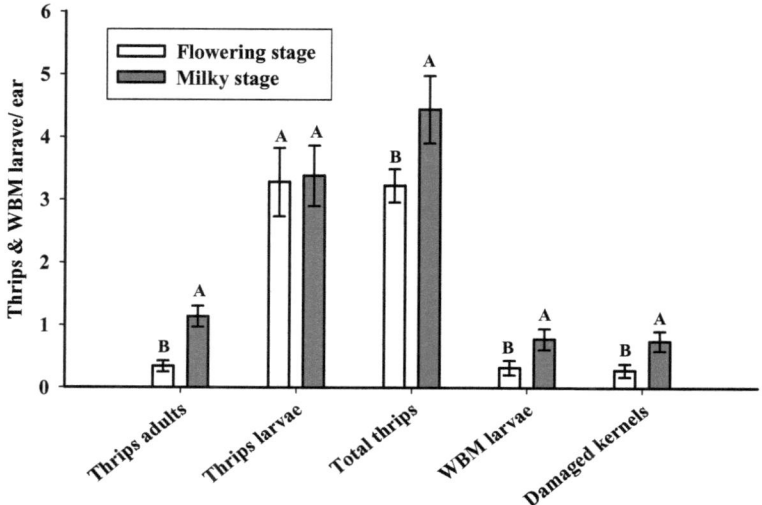

Fig. 40 Mean ± SE of thrips (adults, larvae & total) and wheat midge larvae in two growth stages in 2009. Different letters indicate significant differences.

Correlation between thrips, midges and damaged kernels

There was significant difference ($P= 0.0485$) in damaged kernels by thrips and wheat midges. Treated wheat had lower damaged kernels than control plants. There was a positive correlation coefficient between wheat midge larvae and damaged kernels ($r= +0.56$ and $+0.76$) in GS 65 and 73 stages, respectively; while there was no significantly correlation between total thrips and damaged kernels ($r= +0.121$ and $+0.175$) in both stages (Fig. 39 A& B).

6.5.2.3. 2009

Total thrips (larvae and adults)

There was a significant difference in thrips populations in GS 65 and 73; the significant value was ($P= 0.0030$) in thrips adults and ($P= 0.0484$) in total thrips. While there was no significant difference ($P= 0.891$) in thrips larvae between both stages. The thrips adult were 0.3 in GS 65 and 1.1 / ear in 73, respectively. The corresponding records in thrips larvae were 3.2 and 3.4 / ear. The total thrips were 3.5 and 4.5 / ear in GS 65 and 73, respectively (Fig. 40).

Wheat midge larvae

There was a significant different ($P= 0.0263$) in total midges populations (*S. mosellana* & *C. tritici*) between GS 65 and 73. The total midges were 0.3 and 0.8 larvae/ ear in GS 65 and 73, respectively (Fig. 40).

Damaged kernels by thrips and midges

There was significant difference ($P= 0.0169$) in damaged kernels between growth stages 65 and 73, these values were 0.3 and 0.8 damaged kernels/ ear, respectively (Fig. 40).

6.5.3. Wheat midge larvae population inspected using water traps

6.5.3.1. 2007

Yellow wheat midge larvae were only recorded on GS 75 (1 larva/ trap). *S. mosellana* larvae were significantly higher ($P= 0.039$) on growth stage 85 than other growth stages. The population densities of *S. mosellana* were 6, 4 and 13 midge larvae/ trap at growth stages 75, 83 and 85, respectively. The last larvae were caught on growth stage 87-89 (1 larva/ trap)(Fig. 41).

6.5.3.2. 2008

Populations of wheat midge larvae (*S. mosellana* & *C. tritici*) were significantly higher ($P= 0.0023$) on control than treated. Population density was significantly lower ($P= 0.0353$) on the first two stages (75 and 75-77 (18th & 22nd June)) than other two growth stages (77-79 and 83 (26th & 30th June)). The results indicated that *S. mosellana* & *C. tritici* populations could be divided into two groups; the high populations were recorded on 26th & 30th June and the low populations were recorded on 18th & 22nd June. Mean of low populations of *S. mosellana* & *C. tritici* 1 and 2 midge larvae/ trap in treated and untreated plants. Mean of high populations of both wheat midge larvae were 4 and 12 larvae/ trap in treated and control plants, respectively (Fig. 42).

6.5.3.3. 2009

Yellow wheat midge larvae were only recorded on GS 77 & 87 (1 & 2 larvae/ trap, respectively). Population density of orange wheat midge was significantly higher ($P= 0.028$) on growth stages 83 and 89 than the others. *S. mosellana* numbers were 4 larvae/ trap in both stages. The last WBM larvae were caught on growth stage 89 (Fig. 43).

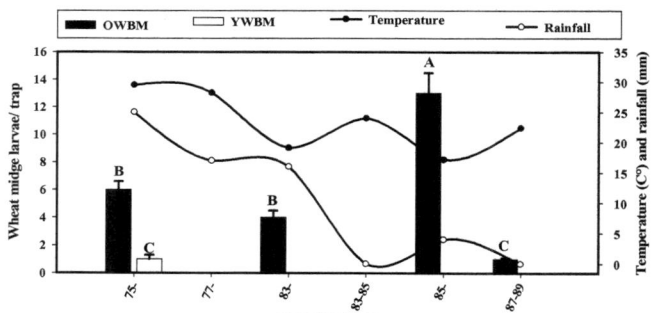

Fig. 41 Mean ± SE of orange and yellow midge larvae by water traps and their relation to temperature and rainfall during winter wheat season 2007.
Different letters indicate significant differences.

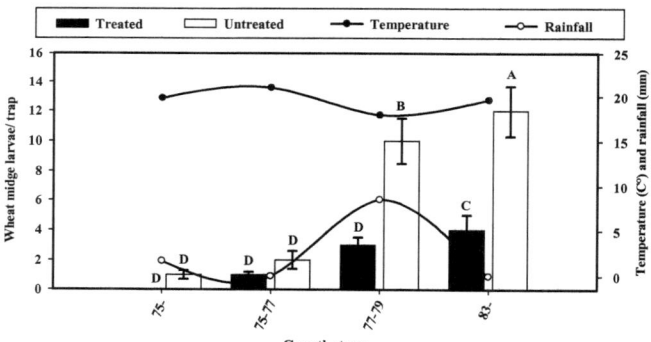

Fig. 42 Mean ± SE of orange and yellow midge larvae catches in treated and untreated plots by water traps and their relation to temperature and rainfall during winter wheat season 2008.
Different letters indicate significant differences.

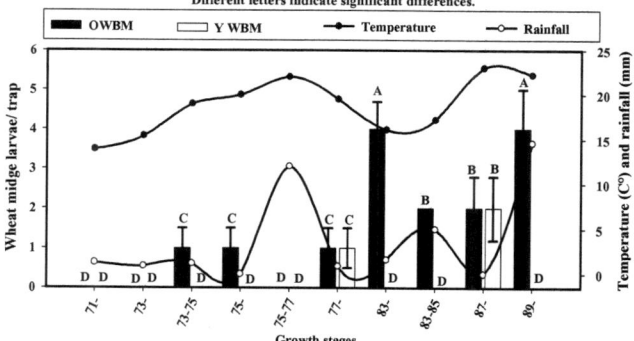

Fig. 43 Mean ± SE of orange and yellow midge larvae by water traps and their relation to temperature and rainfall during winter wheat season 2009.
Different letters indicate significant differences.

6.6. WHEAT MIDGES AND THRIPS EXPERT SYSTEM (WMTES)
6.6.1. General information

An expert system is a computer program, which mimics behaviour of an expert in a particular area of knowledge. Expert systems have been developed and applied in many agricultural fields i.e. diagnose insects and diseases of various crops. Farmers across the world face problems like soil erosion, increasing cost of chemical pesticides, weather damage recovery, the need to spray, mixing and application, yield losses and pest resistance. On the other hand researchers in the field of agriculture are constantly working on new management strategies to promote farm success (Khan *et al.* 2008). In many countries today, farming has become technologically advanced and expert systems are widely used in the field of agriculture. In this way farmers can get expert opinion on their specific problems like selection of most suitable crop variety, diagnosis or identification of livestock disorder, suggestion of tactical decisions throughout production cycle etc. from the expert system. Symptoms of diseases, disorders and pests have due geographical variations. Hence, there is always a need to develop a new expert system for a different geographical region.

Pest management in crops is a highly challenging problem. Globally, annual losses from pests and diseases had increased year after year (Sharma 2001). Sometimes the problem is so severe that loss in crop production is above 50%, and in the case of insect-pest outbreaks, losses are much higher, resulting in complete damage. Different technologies as well as awareness programs are practiced under IPM program for effective, economical and environment friendly control of pests.

The development of an agricultural expert system requires the combined efforts of specialists from many fields of agriculture, and must be developed with the cooperation of the farmers and extension officers who will use them (Chakraborty & Chakrabarti 2008). Expert systems are recognized as an appropriate technology because they address the problem of transferring knowledge and expertise from highly qualified specialists to less knowledgeable personnel (ESICM 1994). Expert Systems can be used by decision makers for predictions, such as on the needs for water, fertilizers and pesticides for a particular crop in the region given the area cultivated with such a crop. This generated information is important for different users: the traders, the exporters, the importers of these materials (Rafea *et al.* 1993; Rafea & Shaalan 1996). Edrees *et al.* (2003) performed an expert system (NEPER) for wheat production dealing with all agricultural practices. This system are verified, validated, and, tested in the wheat fields. There are some expert systems which are used in management systems, for example for aphids (Mann *et al.* 1986; Knight *et al.* 1992; Gonzalez-Andujar *et al.* 1993; Gosselke *et al.* 2001).

The Central Institution for Decision Support Systems in Crop Protection and Crop Production (German acronym ZEPP) was founded in October 1997 on the basis of an administrative agreement of the Federal States. It is dealing with development of simulation method and forecasting insects (Jörg *et al.* 2007) and diseases (Räder *et al.* 2007) on plants to optimize control. Up to date more than 20 met-data -based forecasting models have been developed and introduced into agricultural practice. The results of the forecasting models were improved by the introduction of Geographic Information Systems in the algorithms (Kleinhenz & Roßberg 2000; Kleinhenz 2007; Kleinhenz & Zeuner 2007; Tiedemann & Kleinhenz 2008).

A few expert systems have been reported earlier for wheat insect identification or management. Mann *et al.* (1986) presented a computer-based advisory system for cereal aphid control to winter wheat from early field-cage work on *Sitobion avenae* through a research simulation model. Cereal aphid expert system (CAES) was developed by Gonzalez-Andujar *et al.* (1993); which is designed to provide identification and decision making information to farmers and extension specialists as well as information for educational and research purposes on the main cereal aphid species in Spain. GETLAUS (Freier *et al.* 1996), the latest version of a model for simulating aphid population in dependence on antagonists in wheat (GETLAUS01) was offered by Gosselke *et al.* (2001) for determining cereal aphid population dynamics. Khan *et al.* (2008) discussed a web-based expert system for diagnosis of diseases and pests (Dr. Wheat) for providing an efficient and goal-oriented approach for solving common wheat problems.

Wheat ear insects are perceived as being of major importance, largely because of their variable, yet frequently devastating, effect on farmers' crops. As a result, national and international surveillance schemes have been established, aimed at providing advance warning of pest outbreaks that will allow public and private sector agencies, including farmers for performing agricultural extension services, to make appropriate preparations for insect control (Sivakami & Karthikeyan 2009). In Europe, wheat midge and thrips are two of the most important groups of insect pests (Gaafar *et al.* 2009; Gaafar & Volkmar 2010); some species cause damage directly, through feeding, and indirectly from the fungi infestation. This work is aimed at providing the decision support tools for farmers with rapid access to accurate information that can help them to obtain the threshold to make adequate control decisions.

6.6.2. Model Verification Study (Methodology)

Model verification was done at three sites; two research fields in Halle and Silstedt and one large scale field in Salzmünde, which were selected for detailed study in 2007, 2008 and 2009. The sites were chosen to cover a range of soil types and locations representative of the infested area of central Germany, and to be cover by meteorological stations.

Male WBM numbers were monitored using pheromone traps and ear samples taken to assess the ultimate level of midge larvae infestation in different sites and in two growth stages; flowering (GS 65) and milky (GS 73). White water traps were used to sample the migrated wheat midge larvae to soil in the end of wheat season. For all of these sites the highest catch of male midges in pheromone traps was recorded and compared with infested ears. A correlation analysis was used to investigate any relationship between peak male midge catch and the ultimate level of grain damaged. In general, levels of WBM infestation were relatively correlated with low/ high throughout the monitoring methods and time.

WMTES is Wheat midges and thrips expert system. The observations of variability in trap catch, and how it is related to subsequent infestations, were very relevant when deciding how best to use the traps for WBM risk assessment and were used to develop a decision support model. This model is a distillation of some complicated data obtained over the work in different sites and years but has been framed in terms of what it means for the farmers when using the traps. With this in mind it has been kept as simple and user-friendly as possibly being based on a stepwise decision tree involving yes/no answers to questions (Fig. 44).

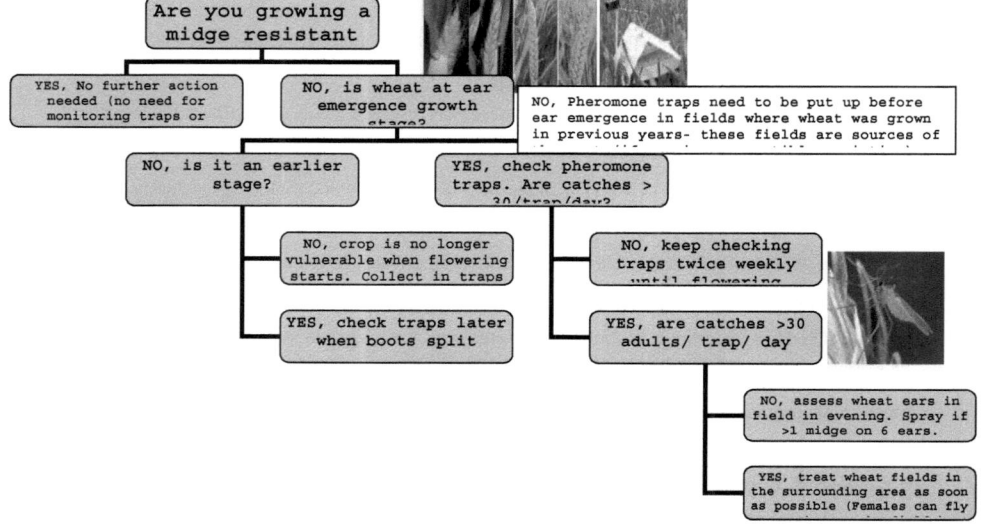

Fig. 44 WBM decision support model (modified after Ellis *et al.* 2009)

6.6.3. Expert System Development Life Cycle and Structure

6.6.3.1. Development life cycle

The first step is creating the knowledge base and the first task in the creation of knowledge base is knowledge acquisition. Knowledge acquisition is considered as one of the most important phases in the expert system development life cycle. Knowledge acquisition is to obtain facts and rules from the domain expert so that the system can draw expert level conclusions (Gonzalez-Diaz *et al.* 2009). Some commonly used approaches of knowledge acquisition are interviews, observations, taking experts through case studies and rule induction by machines. Knowledge acquisition is crucial for the success of an expert system and regarded as a bottleneck in the development of an expert system (Saini *et al.* 2002). After the knowledge acquisition is done, the process of representing that knowledge begins. There are many approaches used for knowledge representation, for example rules, logic expressions and semantic networks. In rule-based expert systems Rules are made on the basis of the hierarchy and these rules lead to proper treatment that the user has to use.

The domain must be compact and well organized. The quality of knowledge highly influences the quality of expert system (Suo & Shi 2008). The first step in the development of any expert system is problem identification. The problem here is a diagnostic problem aimed to identify ailments in the wheat using symptoms of insect pests. The problems occur frequently and the consequences on farmer's financial status are enormous. The demand for help is increasing rapidly. Diagnosis or diagnostic problem solving is the process of understanding what is wrong in a particular situation. Thus gathering of information and then interpreting the gathered information for determining what is wrong are of central importance in diagnostic problem solving (Lucus 1997).

6.6.3.2. System Structure

Figure (45) shows typical expert system structure we have created. Each of these blocks is explained below.

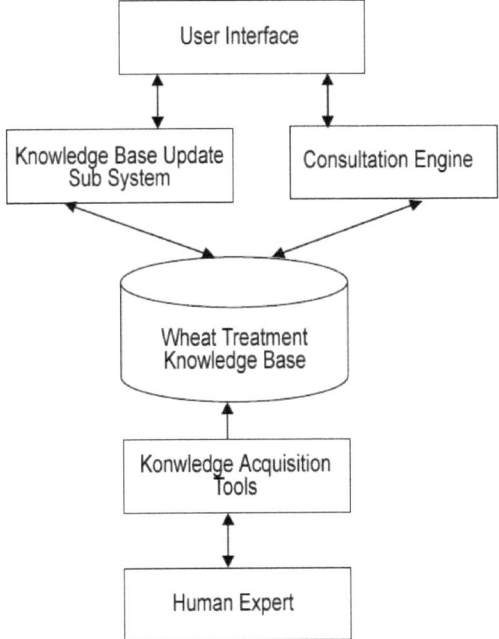

Fig. 45 Expert system structure of wheat ear insects (modified after Khan *et al.* 2008)

6.6.3.3 User Interface
This is the interface the end user will use to interact with the system by providing parameters to it and having recommendations and consultation results out of it.

- **Knowledge Base Update Sub System**
  As we know the utmost drawback of using expert system is that it has fixed knowledge base. If this base is not updated periodically, the results of consultation by time will be out of date. Thus, we developed this sub system to have the ability to update and enhance the knowledge base at any time easily and smoothly.

- **Consultation Engine**
  Consultation Engine is the communication channel between the end user and the system; this is where user submits his consultation. Engine has 2 operation modes one is system wizard and the other is manual entry, and we will

- **Wheat Treatment Knowledge Base**
  This is the heart and the core of the system where it holds all the knowledge that we process to give the right decision to the user.

- **Knowledge Acquisition Tools**
  This is the ways we acquire knowledge from different sources and save it in the knowledge base.

- **Human Expert**
  Everything in the end must return back to humans without the help of human expert we can not by any means have computerized expert system.

6.6.3.4 System User Interface
Our system has 3 main modules:
1- System main data entry module
2- Knowledge base update module
3- Consultation engine module

6.6.3.4.1 System Main Data Entry Module: System Main Screen (Fig. 46A)
Main Data Menu: (Fig. 46B)

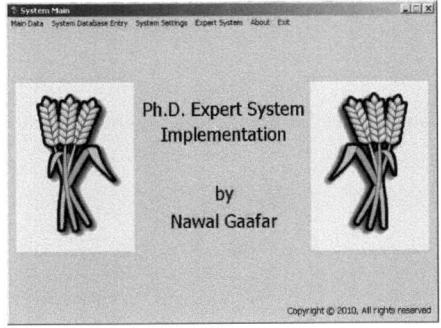
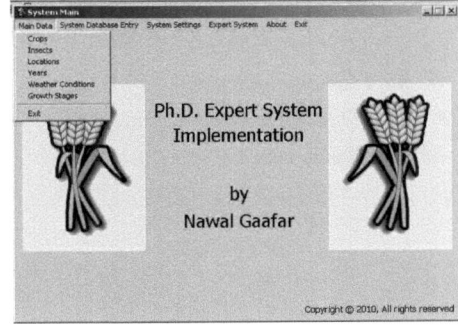

Fig. 46A System Main Data Entry Module   **Fig. 46B** Main Data Menu

Crops Window

Here we can add, update and delete any kind of crops that we are dealing with now or may be in need to deal with in the future (Fig. 47A).

Insects Window

From this window we can add, update and delete any kinds of insects that we are dealing with now or may be in need to deal with in the future (Fig. 47B).

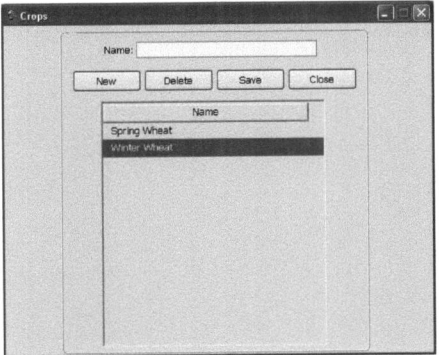
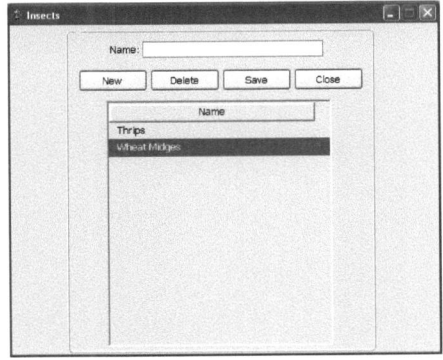

Fig. 47A Crops window   **Fig. 47B** Insects window

Locations Window

From this window we can add, update and delete any study locations that we are using now or may use in the future (Fig. 48A).

Weather Conditions Window

From this window we can add, update and delete any weather conditions that may be affected either on crop or insects (Fig. 48B).

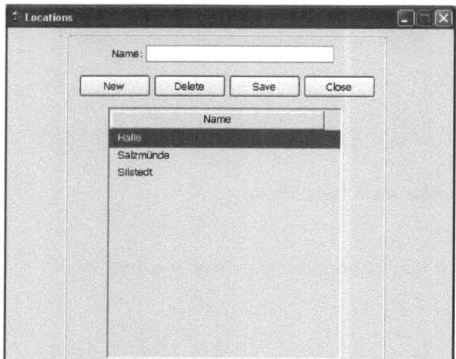
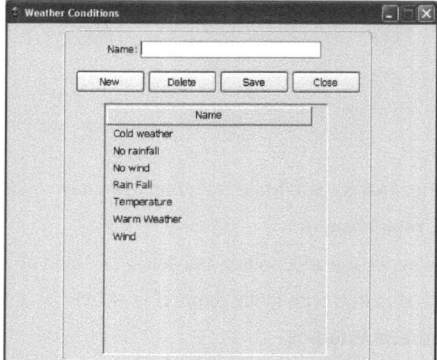

Fig. 48A Locations window   **Fig. 48B** Weather conditions window

Growth Stages Window: From this window we can add, update and delete any growth stages that we are interesting to study the population dynamic of insects (Fig. 49).

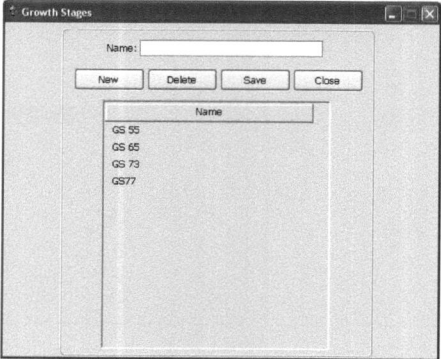

Fig. 49 growth stages window

6.6.3.4.2 Knowledge Base Update Module

In this window we will be able to modify and update the knowledge base we have to be concurrent with the latest researches and results we got from different data acquisition techniques (Fig. 50).

Fig. 50 Knowledge Base Update Module

6.6.3.4.3. Consultation Engine

System Settings: In this window we can change the consultation engine setting by choosing the defaults of the system parameters and even saving it permanently in the system (if you click on save) or just change it for the current consultation session (if you click on apply) (Fig. 51)

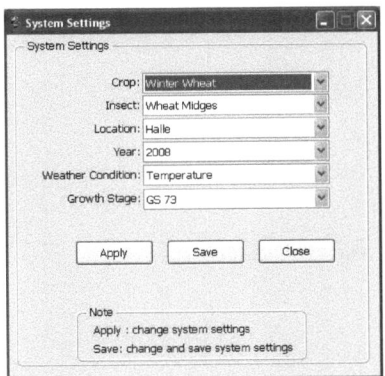

Fig. 51 System Settings

- **Consultation Engine Menu**

    This menu contains menu items for System Wizard, Pheromone Traps, Evaluation of Wheat Ears and Water Traps

- **System Wizard**

In system wizard the system will keep asking questions and get answers from user till it has all the required information to give right decision for the user the following figures will appear to the user in order (Fig. 52A, B, C, D, E, F, G, H and I).

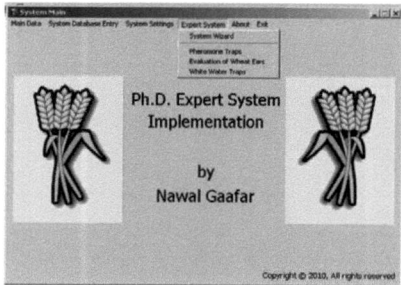

Fig. 52A System Wizard

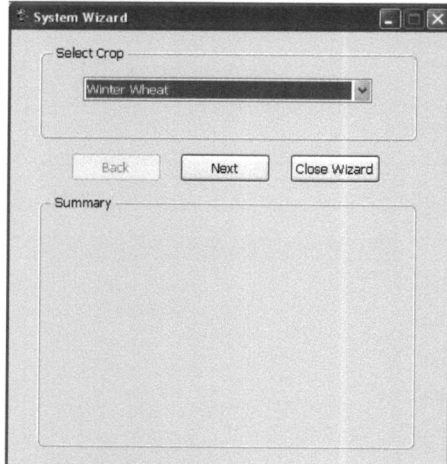 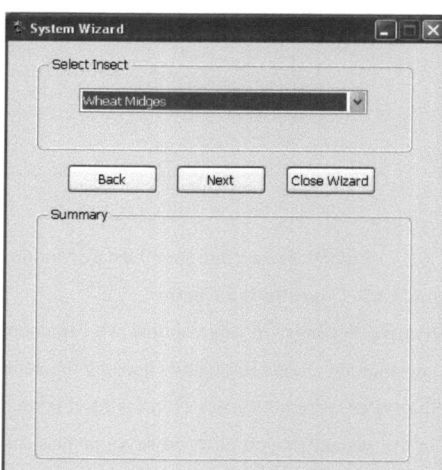

Fig. 52B Select crop   **Fig. 52C** Select insect

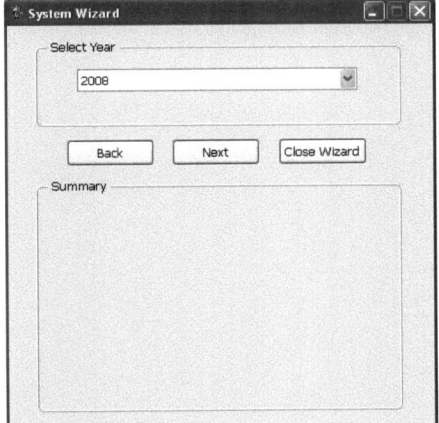
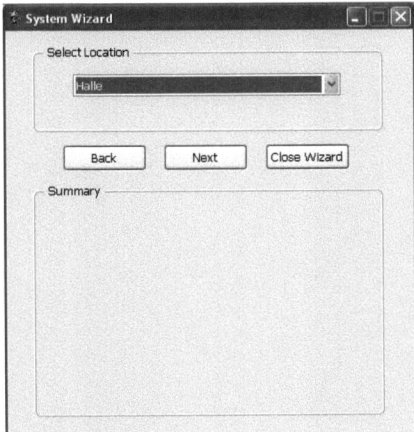

Fig. 52D Select year

Fig. 52E Select location

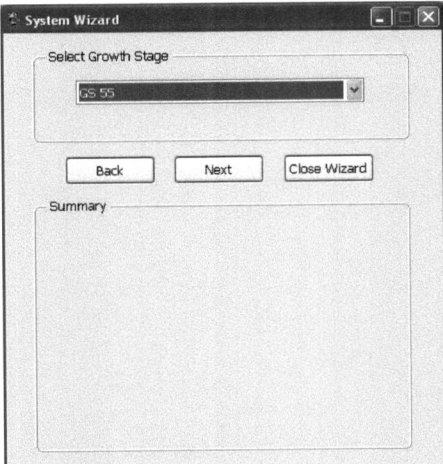
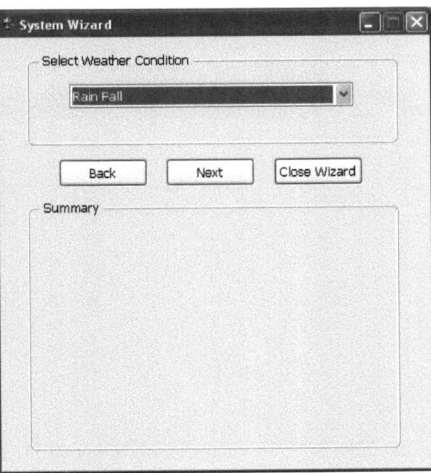

Fig. 52F Select growth stage

Fig. 52G Select the weather conditions

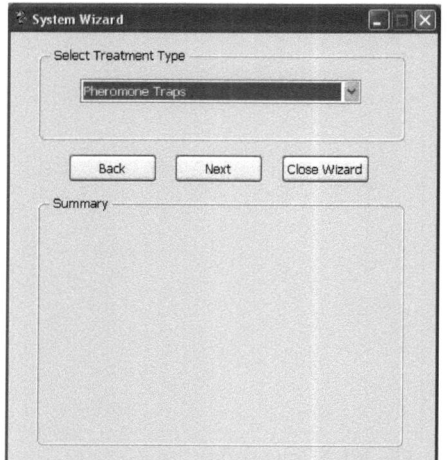
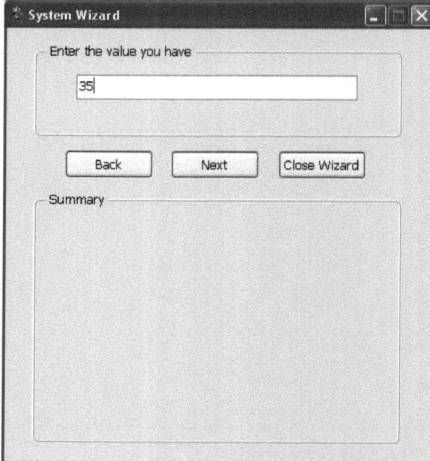

Fig. 52H Select treatment type   **Fig. 52I** Enter the value you have

Summary window after gathering all the required information from the user (Fig. 53A) This is an example for the consultation result out of the system (Fig. 53B).

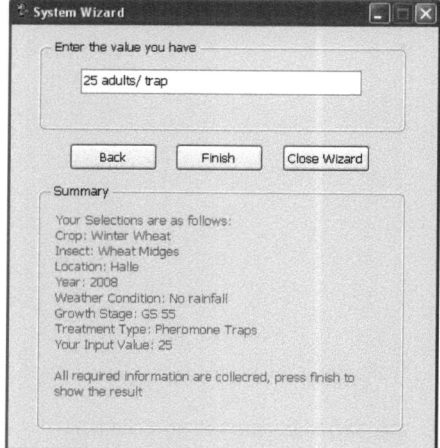
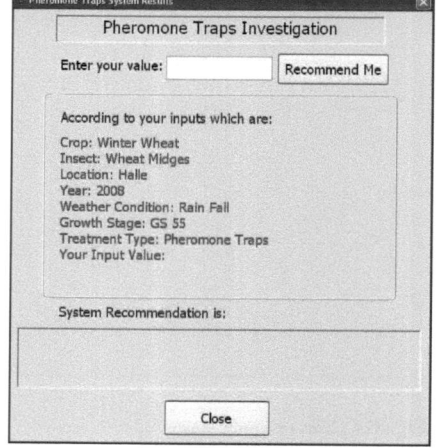

Fig. 53A All the required information   **Fig. 53B** The consultation result out of the system

Pheromone Traps: In this window system will use the defaults assigned in system settings in the consultation where user will only submit the OWBM value and click on recommend me and the system will process the value and give recommendation to user (Fig. 54 A&B)

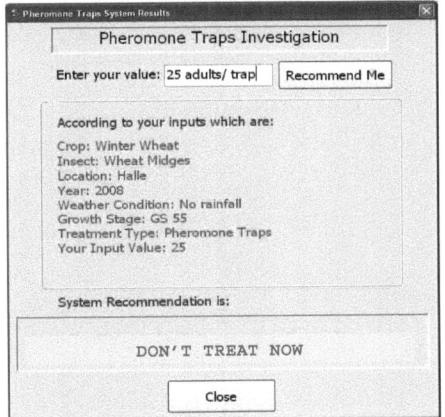 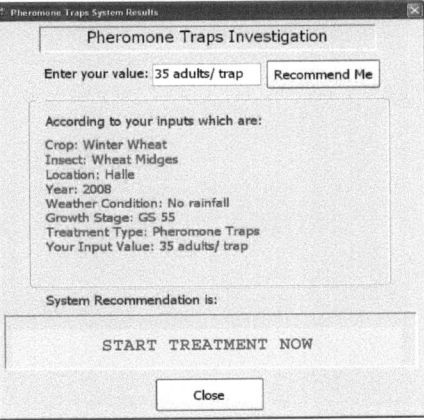

Fig. 54A Recommendation without treatment   **Fig. 54B** Recommendation with treatment

Evaluation of Wheat Ears (Midges): Here, this system will use the defaults assigned in system settings in the consultation where the user will only submit midge larvae value and click on recommend me and the system will process the value and give recommendation (Fig. 55 A&B).

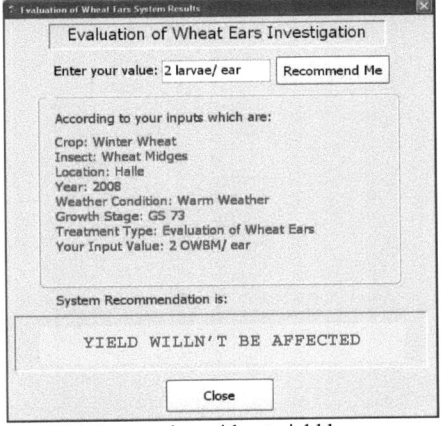 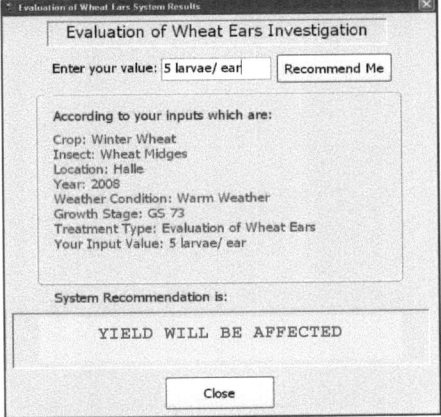

Fig. 55A Expectation without yield losses   **Fig. 55B** Expectation with yield losses

Evaluation of Wheat Ears (Thrips): This system will use the defaults assigned in system settings in the consultation where the user will only submit thrips value and click on recommend me and the system will process the value and give recommendation to the user (Fig. 56 A&B).

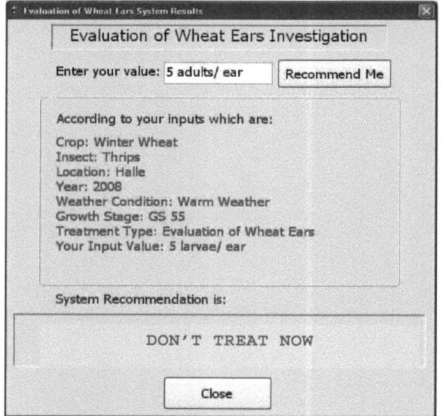 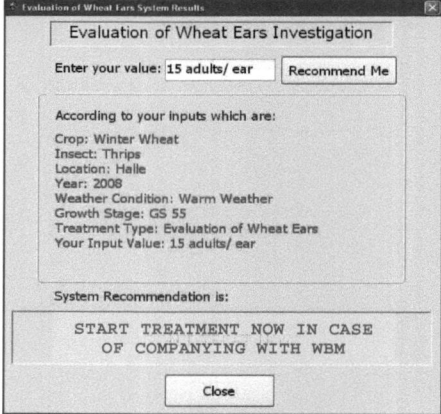

Fig. 56A Recommendation without treatment **Fig. 56B** Recommendation with treatment

White Water Traps: In this window, the system will use the defaults assigned in system settings in the consultation where the user will only submit the midge larvae value and click on recommend me and the system will process the value and give recommendation (Fig. 57 A&B).

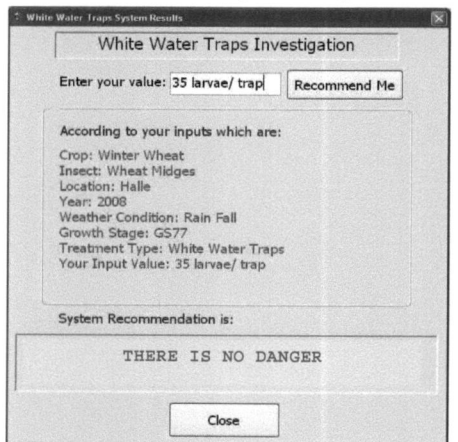 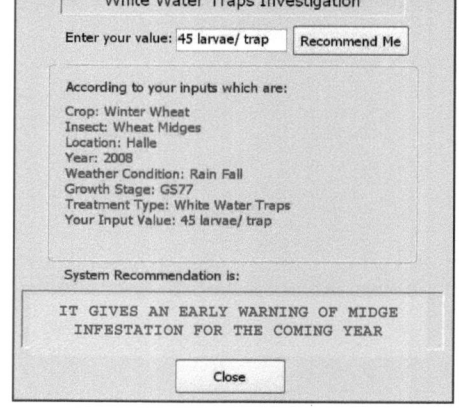

Fig. 57A Recommendation without danger **Fig. 57B** Recommendation with warning

About

This is the about and copyright of the system (Fig. 58)

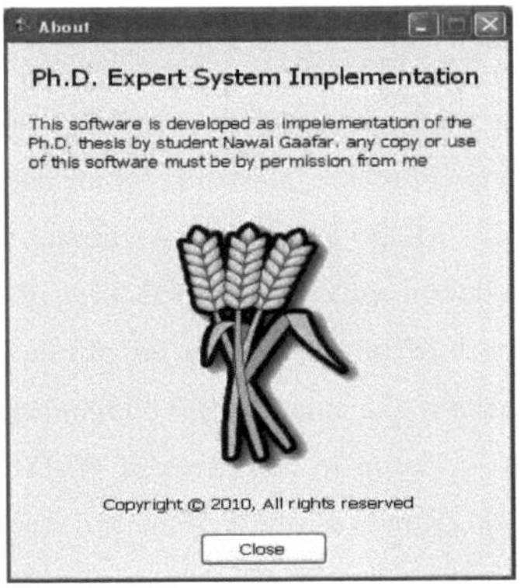

Fig. 58 The copyright of this system

6.6.4. Testing and validation

WMTES field trials (based on crop samples) during 2007, 2008 and 2009 seasons, were done for the samples received from three different regions. As it is unwise to implement any ES from day one after completion of its development, parallel consultation from the IPM experts was found useful in improvement and validation of ES results. Also, real world ES need testing and validation in the real world environment i.e. field testing. Comparison of results produced from system as well as suggested by IPM experts has been used to improve the quality of inferences and consultation. Feedback forms have been used for preparation of validation case sheets.

6.6.5. What should be concluded?

Pheromone traps only monitor male midge populations whereas it is female WBM that lay eggs from which the damaging stage of the pest emerges. Therefore, it is not possible to set a simple trap catch threshold above which economic damage occurs and below which it does not. A decision support model that can be used by farmers was developed using a stepwise decision tree involving yes/no answers to questions (Fig. 44). When growing a susceptible wheat variety pheromone traps need to be put out before ear emergence in fields where wheat was grown in previous years and provide a source of the pest. These traps should be monitored daily or at least twice weekly during the susceptible growth stage. When trap catches exceed threshold (30 midges per/ day) crop inspections provide additional information to help decide whether to treat a field. It is advantageous that the pheromone traps are so sensitive and catch as many WBM as possible because they provide an early warning of midge flight during ear heading time and the suitable weather conditions, thus avoiding situations in which insecticide sprays are applied too late when they are needed. The threshold of midge larvae infestation is 3-4 larvae/ ear, where as in water traps, used to sample the migrated midge larvae is 30-40 larvae/ trap. The later should be also monitored carefully after the heavy rain, particularly at late milky stage; it gives an early caution of midge infestation for the coming year, especially intend cultivating wheat after wheat.

Midge infestation was higher in 2008 than in 2007 and 2009. The low levels of midge infestation hindered the verification of the decision flow tree as it was not possible to examine the impact of midge catches on infested ears. There can be some confidence in the proposed threshold of greater than 30 midges/ trap/ day to indicate a need to inspect crops for the pest.

The main objectives of WMTES have been met in allowing better data provision at all levels. However, the system can easily be changed and updated to meet new demands. For example, the current system automatically produces bulletins. This function could be developed further to produce automatically customized pest reports to include the interpretation of current pheromone or water traps data and to provide pertinent advice to growers and/or advisors. These could include comparisons between ear insect numbers in previous years and changes in the numbers between dates for specific regions, or a more forecast for weather conditions on the population dynamic of ear insect in any year.

If the research requirements should change, the present set of data extraction programs could be augmented by additional routines, for example, for summing the total wheat midges or thrips within a particular group of species. Other decision support systems could also be added to the system, as and when information and expertise are available to develop them. The use of a database system, such as WMTES, is not limited to these pests but is equally applicable wherever forecasts are made on the basis of large amounts of data collected by a monitoring scheme.

Although systems of this sort allow greater use to be made of monitoring data, there are still several constraints to their full use. In Germany, as in many other countries, there is increasing pressure for public sector institutions to become more commercial, thus reducing the free exchange of information. Institutions are encouraged to charge for the information they supply and to register the intellectual property rights of any potentially useful product. These constraints on the free interchange of information have led to a reduction in the distribution and use of potentially valuable crop protection information. It is important that such constraints are accounted for in the specification of similar database projects.

In general, the current expert system has improved understanding of the WBM problem. Improving risk prediction: the decision flow chart proposes thresholds to help predict the need for insecticide treatment against WBM. The verification study suggested that these are a good basis for risk management. However, thresholds are based on data from a limited number of sites and years; further work is required to confirm the initial findings and improve the precision with which it is possible to predict the risk of pest attack. Risk of damage is also primarily dependent upon the coincidence among midge activity, the susceptible stage and weather conditions. Being able to predict the likely timing of the susceptible growth stage in relation to midge emergence would be a significant development, and help to limit unnecessary insecticide treatment.

WMTES integrates hypermedia advantages for information and knowledge supporting the ES recommendations for their better utility and increased dissemination capability through user friendly interfaces, as reported by Carrascal *et al* (1995) for using hypermedia in knowledge and information transfer in agriculture. Estimation of pest activity using computational techniques has to incorporate variations from exact activity in the field. Practically, the relaxations given by the experts lead to acceptance in recommendations in up to 95% cases, except in the cases where differences in results are more than $\pm 10\%$. A knowledge base developed for a specific region in a problem can be tuned to make it more appropriate for other regions.

Much of the power and flexibility of expert systems are due to the fact that the knowledge base is separated from the inference mechanism (Waterman 1985; Rauscher 1990). The knowledge base can be modified without interfering with the operation of the system or the performance of other rules. In this way the program is now being enriched with rules for other insects, crops and sites, as well as by using of the user's feedback to improve the system. It is as important as to develop the system to maintain it. This task will be carried out for the extension services. WMTES is designed to enhance the quality and availability of knowledge required by decision makers in wheat insect management. It depends on a knowledge base that contains all the knowledge required to give useful, accurate and adequate consultations to wheat farmers.

The expert system will be used in training new 'experts'. It will allow less experienced users to examine the reasoning process of an expert, to improve their understanding of how one takes control decision and to learn how to approach different situations to take the adequate decision. Agricultural extension services require more effective ways of handling, communicating, and using information. The program for the control of wheat ear insects is an example of one way that the small expert system technology can be successfully applied to the solution of every day problems in plant protection.

7. DISCUSSION

There was no coincidence in 2007 between wheat midge activity and susceptible stages of wheat. Therefore wheat plants had escaped from midge infestation, because heading phase was earlier ca. 10 days and the hibernated midges emerged later due to warm and dry weather (0.1mm) in April. The results indicated that a much greater risk of damage from a given wheat midge adult population would be anticipated when settled cyclonic weather coincided with adult midge flight and ear emergence, as occurred in Halle 2008, Salzmünde 2008 and Silstedt 2007 and 2008 based on weather conditions. On the other hand, the situation in 2009 is considered a moderate level between 2007 and 2008.

7.1. Silstedt site

Pheromone trap catches were affected with susceptible stages and weather conditions as mentioned above. In 2009, there was one peak in the early of season (GS 43), that returned to heavy rain (6mm) at early of May, followed by warm still temperature which affect emergence of adult midges earlier. These results are consistent with Oakley *et al.* (2005) and Bruce *et al.* (2007), and Gaafar & Volkmar (2009) who stated that numbers of wheat midges do give an accurate prediction of risk. Rainfall and temperatures proved to be accurate in predicting adult midge activity, as temperature in spring apparently had effect on subsequent development.

Independent of the different levels of thrips and midge larvae attack in all experimental years, Türkis, Tommi and Potenzial proved the most susceptible varieties to WBM, while Brompton, Skalmeje, Robigus, Welford and Glasgow showed a clear resistance reaction. These results are similar with Volkmar *et al.* (2008, 2009), Gaafar *et al.* (2009, 2011) who studied some wheat varieties in Germany for their susceptibility to thrips and WBM infestations.

Wheat midge populations were significantly positively correlated with the number of damaged ears among varieties in the studied years. These findings coincide with that concluded by Olfert *et al.* (1985) and Smith & Lamb (2001) who mentioned that such a strong correlation was expected because midges prefer to oviposit the eggs in wheat ears in the flowering stage, and when hatched, quickly move into the ear and damage kernels in the milky stage.

Variety trials showed consistently low wheat midge larval infestations on the resistant varieties such as Brompton, Skalmeje, Robigus, Welford and Glasgow. In contrast, infestation levels on other varieties varied between seasons such Boomer, Potenzial, Global and Tommi. Although variety such as Potenzial has *wm1* marker, but it is not resistant to orange wheat midges and its infestation vary from year to year. On the other hand, Skalmeje variety has also this *wm1* marker, and it is a resistant variety, therefore, midge infestation was very low as confirmed by Schliephake (2009, personal communication); Blake *et al.* (2010). The low densities of wheat midge on resistant wheat in this study are consistent with those found in previous studies conducted with different objectives (Lamb *et al.* 2000a; Smith *et al.* 2004). Similar results were obtained by (Smith *et al.* 2007), pointing that wheat midge developing on resistant wheat was always very low compared with that of larvae developing on susceptible wheat varieties. They also mentioned that small numbers of *S. mosellana* matured larvae in each wheat variety carrying the *Sm1* gene for antibiosis resistance against this insect. Synchronicity between the susceptible ear emergence stage of the crop and the peak of WBM flight activity was another key factor in determining larval infestation levels. The resistant varieties had the lowest levels of WBM larval infestation as expected. This result confirms the importance of monitoring pest numbers in order to make the decision for insecticide treatment.

This may be due to that the ancestors in these varieties evolved a defence mechanism. Explanations for such a mechanism: The most resistant wheat varieties had a higher constitutive level and a more rapid induction of ferulic acid than susceptible varieties, which increased the mortality of newly hatched larvae. Analysis of phenolic acids in grain samples showed that levels of ferulic acid were higher in infested grains of some varieties compared to uninfested grains. Levels of *p*-coumaric acid were greater in the infested than in the uninfested samples of all the tested varieties indicating that WBM damage is inducing production of this acid in the seed as reported by Ellis *et al.* (2009). This suggests that there might be another mechanism of WBM resistance (Smith *et al.* 2007).

Understanding the biochemical basis of resistance: although it is clear that the *Sm1* gene is responsible for resistance, as in Canada (Smith *et al.* 2007), the mechanism of resistance is still not understood. Canadian research suggested a correlation between increased levels of ferulic acid and resistance. Further investigation is required to help future breeding programs (Ellis *et al.* 2009). Another demonstration by Ding *et al.* (2000) who mentioned that few wheat varieties have a high level of antibiotic resistance to the larvae, which suppresses their development; nearly all larvae develop successfully on susceptible wheat varieties (Thomas *et al* 2005).

There were less thrips or midges in the damaged kernels of some varieties than others in the studied years. Some wheat varieties also have evolved a defence mechanism that deters oviposition by the wheat midge as mentioned by Berzonsky *et al.* (2003). These discrepancies might have been a result of speed ripening time as reported by Elliott *et al.* (2000). The wheat midge has evolved preferences for ovipositing at particular developmental stages of its host. This may have been sufficient to make some varieties less favourable for oviposition (Lamb *et al.* 2001). Such varieties are recommended to be cultivated in the next year.

Glasgow and Skalmeje varieties had the highest number of *S. mosellana* larvae in water traps in 2009; although they are resistant varieties to OWBM. This can explained by that the resistant varieties have a defence system to larvae maturing, but don't have such system to oviposition. By this way, there was no deterrence for egg laying but there was an antibiotic resistance which prevents the normal larvae growth, therefore the larvae in these varieties were migrated very early in the first instars as noticed in our work. This result was corresponded with that obtained by Simth *et al.* (2007), who mentioned that mean mass of larvae from resistant wheat was significantly less than that of larvae from susceptible wheat.

Resistant wheat carrying the *Sm1* gene for antibiosis to wheat midge is highly effective in preventing the development of larvae. However, a small proportion of larvae were capable of maturation on resistant wheat and surviving to adult emergence, although at a lower rate than larvae did on susceptible wheat. Another possibility is that occasional larvae survive because of variability in the expression of the *Sm1* gene in wheat plants resulting from environmental influences as reported by Lamb *et al.* (2001). If a lower degree of infestation is predicted, producers may stick to their plans to grow wheat, but may choose less susceptible and early heading wheat varieties to avoid high midge populations during heading stage.

7.2. Halle site (winter wheat)

Field-trapping experiments with pheromone traps of *S. mosellana*, demonstrated that they were highly attractive to males and caught very few non-target organisms. These results are consistent with those recorded by Oakley *et al.* (2005), Bruce *et al.* (2007) and Volkmar *et al.* (2008). Weather conditions (rainfalll and temperature) proved to be accurate in predicting orange midge activity, as temperature in spring apparently had effect on subsequent development. As a result, these findings confirm the utility of pheromone traps as a decision-making tool for management of WBM populations in winter wheat fields.

There were significant differences in the number of thrips and WBM recorded in wheat ears among winter wheat varieties in the successive years. Numbers of thrips were higher in

Ritmo in 2008 and Kontrast in 2009. The lowest thrips populations were recorded in Thuareg in 2008 and Robigus in 2009. The midge larvae numbers were higher in Michigan Amber in 2008 and 2009, while the lowest WBM numbers were observed in varieties Boomer and Akratos. These results are similar with Volkmar *et al.* (2008), Gaafar *et al.* (2009) and Schröder (2009), who studied some wheat varieties in Germany for their susceptibility of thrips and wheat midges.

The damaged ears were significantly positively correlated with wheat midge numbers among varieties and between years. A strong correlation was found between damaged ears and the number of midges per ear. The same results were also reported by Olfert *et al.* (1985) and Smith & Lamb (2001), pointing that such a strong correlation was expected because midges prefer to oviposit the eggs in wheat ears in the flowering stage, which hatch and then quickly damage the full kernel in the milky stage.

There were some wheat varieties which showed no infestation with WBM larvae in the studied years including Vitco, Enorm, Marzurka, Capo, Welford and Robigus (the first four varieties are early heading and the last two are resistant to wheat midge). It is proposed that ancestral lines of these six varieties may have evolved a defence system which reduced or prevented egg-laying on spikes, based on the following observations: 1) larvae perform well in some varieties and so genotypes with this trait would not select for reduced oviposition by wheat midge; 2) wheat midge females show a much reduced preference for the same varieties when spikes emerge; 3) reduced preference is a stable trait as stated by Smith & Lamb (2001) or characteristic of some plant genotypes as mentioned by Ganehiarachchi & Harris (2007); 4) the overall reduced preference for those varieties is associated with a larger proportion of eggs laid on the rachis than is the case of other varieties. Eggs laid on the less preferred sites, such as the rachis and exposed surfaces of glumes, are likely to be more exposed to desiccation and parasitism than those laid on the inner surface of glumes or in florets. Furthermore, larvae that hatch from eggs laid on the rachis must travel a greater distance to reach their feeding site. Therefore, it is likely that deterrence of oviposition on glumes would not only reduce egg density but also the subsequent performance of wheat midge eggs and newly hatched larvae as reported by Lamb *et al.* (2001; 2002). Identifiable morphological traits have been associated with a similar oviposition deterrence that occurs in some wheat varieties such as Capo (with hair) or Vitco which is an early heading variety which escape from WBM infestation as stated in other studies by Gaafar *et al.* (2009). The results of other researchers such as Birkett *et al.* (2004) point to a role of plant chemistry such as volatiles in host finding and selection. Another demonstration by Ding *et al.* (2000) mentioned that few wheat varieties have a high level of antibiotic resistance to the larvae and other have *Sm1* (McKenzie *et al.* 2002), which suppresses midge growth and development. Nearly all larvae develop successfully on susceptible wheat varieties. The most resistant wheat varieties had a higher constitutive level and a more rapid induction of ferulic acid than susceptible varieties, which increased the mortality of newly hatched larvae.

There were less thrips or midges in the damaged kernels of some varieties than others in the successive years. Some wheat varieties also have evolved a defence mechanism that deters oviposition by the wheat midge as mentioned by Berzonsky *et al.* (2003). These discrepancies may have been a result of fast ripening time as reported by Elliott *et al.* (2000). The wheat midge has evolved preferences for ovipositing at particular developmental stages of its host. This may have been sufficient to make some varieties less favourable for oviposition. Such varieties are recommended to be cultivated in the next year.

7.3. Comparison of Halle & Silstedt in 2008 season

There were significant differences in thrips and WBM among varieties in both sites Halle and Silstedt. Numbers of thrips were higher in the varieties Türkis and Welford in Halle and Türkis and Anthus in Silstedt. The least thrips numbers were recorded in varieties Potenzial and Boomer in Halle and Robigus and Potenzial in Silstedt. This result is similar with Volkmar *et al.* (2008, 2009), studying some wheat varieties in Germany for their susceptibility of midges.

There were significant differences in the midge larvae. Their numbers were higher in varieties Tommi and Potenzial in Halle and Türkis and Dekan in Silstedt, while the least WBM numbers were observed in varieties Anthus, Welford and Robigus in both Halle and Silstedt.

The ears damaged were significantly positively correlated with wheat midge numbers among varieties and between both sites. A strong correlation was found between damaged ears and number of midge per ear, same results were also reported by Olfert *et al.* (1985) and Smith & Lamb (2001), who mentioned that such a strong correlation was expected because midges do not prefer to oviposit in wheat ears that are already infested.

There were more thrips or midges and damaged kernels in some varieties than others in the two sites. Some wheat varieties also have evolved a defence mechanism that deters oviposition by the wheat midge as mentioned by Berzonsky *et al.* (2003). These discrepancies may have been a result of speed ripening time as reported by Elliott *et al.* (2000). The wheat midge has evolved preferences for ovipositing at particular developmental stages of its host. This may have been sufficient to make some varieties less favourable for oviposition, such varieties are recommended for cultivation in the next year. If a lower degree of infestation is predicted, producers may stick to their plans to grow wheat, but may choose a less susceptible wheat variety and plant early to avoid high midge populations during heading.

7.4. Halle site (spring wheat)

Catches of orange wheat midges in pheromone traps were coincident with susceptible growth stages of wheat (GS 47-65). There was a strong correlation between the peaks of pheromone trap catches and weather conditions. These results are similar with those obtained by Volkmar *et al.* (2008) and Gaafar & Volkmar (2009). If trap catches indicate that a significant number of adults may occur in the next crop year, then adult populations and weather during the susceptible stage of the wheat crop need to be closely monitored. A correlation between midge catches and weather conditions was obtained; this gives base for control decision making.

There was a significant difference in the number of midge larvae as well for the thrips between varieties. Thrips and midges were higher in Triso than in Sakha 93. Therefore, Triso was more affected than Sakha 93 in the number of damaged kernels. There was a correlation between WBM numbers and damaged kernels. These differences between the two varieties may be depending on different morphological traits of ear (with or without hair). The second explanation for differences is that some wheat genotypes express oviposition deterrence which can cause a lower infestation of some genotypes as reported by Lamb *et al.* (2002, 2003), Olfert *et al.* (2009) and Blake *et al.* (2010). A third reason may be due to different concentrations of ferulic acid as confirmed by Abdel-Aal *et al.* (2001), pointing that ferulic acid in the wheat kernel, was found to differ significantly in the mature grain of six spring wheat varieties known to have a range of tolerance to *S. mosellana*.

There was a difference between the studied years in midges as well thrips infestation. Midge infestation was higher in 2008 than 2009 season. This should be due to the influence of the environmental conditions. In case of dry stress during May and June, larvae remain dormant until the next year as confirmed by Elliott & Mann (1996).

Orange wheat midge larvae in white water traps were higher in 2009 than 2008, while yellow wheat midge larvae were vice versa. Usually, WBM larvae wander from wheat ears at dough stage and burrow in the soil to a depth of 0–6 cm as stated by Doane & Olfert (2008). The high catches of WBM in the water traps on some dates may be attributed to weather conditions, especially rainfall, which had a direct effect to help midge larvae migrate to the soil. Similar results were recorded by Gaafar et al. (2011) in winter wheat, stating that catch densities had increased in the dates after heavy rainfall directly.

Yield in Triso variety was higher than in Sakha 93, although the later had lower midges and thrips infestation. This may be due to the adaptation of Triso as German variety in contrast to Sakha 93 which was brought from Egypt as drought resistant variety or may be attributed to the number of tillers which were more in Triso than Sakha 93. Grain yield and quality parameters were affected by the genotype of the varieties or its parent effects (The parents of Sakha 93 had low wheat yield); similar results were recorded by Cesevičienė et al. (2009), who stated that different genotypes of the varieties affected the wheat yield.

WBM can cause damage to spring wheat that reaches economic levels, as well as in winter wheat. The results suggest that a trapping program based on pheromone and water traps should be the best way of sampling wheat midges in the early and end of season.

7.5. Salzmünde site

Large variations in Orange midge numbers caught in the pheromone traps and in timing of peak catches were found between years (ca. fivefold) in farm scale studies undertaken in Salzmünde. This suggests that it is more useful for farmers to put traps in neighbouring fields which were cultivated wheat in the previous year. In general, in 2008 and 2009 the peak of midge flight synchronized with the susceptible stage of the crop, it was more adequately in 2008 than in 2009, damage levels tended to be higher in 2008 than in 2009, because there was correlation between total numbers of males caught during the susceptible period and infestation as confirmed by Ellis et al. (2009). Pheromone traps were very valuable in indicating midge emergence and for decision making. This is a significant benefit along with other systems for monitoring wheat midges as mentioned by Gaafar & Volkmar (2009) and Gaafar et al. (2011).

The peaks of pheromone trap catches for the whole season occurred when the wheat was past the susceptible growth stage for example 2007, but for setting the economic threshold the peak catch during the susceptible period is more relevant. Routine use of this monitoring method should eliminate most unnecessary applications of insecticides, and help assure that the benefits of insecticide applications exceed the cost as confirmed by Volkmar & Gaafar (2010), who presented how can monitor OWBM by pheromone traps in practice wheat fields in Salzmünde.

Levels of midge infestation were higher in 2008 than in 2009 in both methods; evaluation of ear insects and white water traps. This meant that there was a good correlation between pheromone trap catches and midge infestation. Therefore, chemical control was applied in 2008 and not applied in 2007 or 2009 in case of low levels of midge infestation in 2007 and 2009. As a consequence, although the pheromone traps indicated that OWBM had emerged, the time of arrival of many of the egg-laying females in the crop was not synchronized with the susceptible growth stage. This explains why there was a poorer correlation between pheromone trap catches and subsequent infestation in 2007 or 2009 than in 2008. Similar results were recorded by Ellis *et al.* (2009), who reported that difference in weather condition between 2004 and 2005 had direct affected on wheat midge populations.

Although levels of midge infestation were generally higher in 2008 than in 2007 and 2009, there is evidence suggesting that the proposed thresholds for control decision are a good basis for prediction the risk of midge attack. If cumulative trap catches exceed 30 midges/ trap/ day during the susceptible stages (GS 47-65), then this indicates an economic risk to the wheat crop and an insecticide application may be necessary. Our results are consistent with those observed by Oakley (2008), Ellis *et al.* (2009), who determined the threshold as 30 or more midges caught in a pheromone trap per day. As well as for ear evaluation, three or four maggots per ear will destroy the kernels in that ear. Similar results were found by Olfert *et al.* (1985, 2004), Oakley *et al.* (2005) and Ellis *et al.* (2009) in Canada and UK, they confirmed that if one or more adult midges are observed for every 4-5 heads or 3-4 midge larvae/ ear. Chemical control may be required when reach >1 OWBM adults/ 6 ears (Ellis *et al.* 2009) and >1 YWBM adults/ 3 ears (Volkmar & Wetzel 1989) at GS 55. The threshold for thrips was 10 thrips/ ear at heading stage (GS 55) and 25 thrips/ ear at flowering stage (GS 65), especially when thrips are companied with wheat midges at both stages. However, Wetzel & Freier (1981) mentioned that threshold of thrips was 5-10 thrips/ ear at flowering stage. This means that the threshold of thrips was higher in our work during 2007 to 2009 than in 1980s, also as confirmed by Volkmar (1988) who stated this value was 3 thrips/ ear. Pheromone traps, ear inspection and midges captured indicate the class of risk, while the threshold indicates the need for control.

7. 6. Expert system

The main objectives of WMTES have been met in allowing better data provision at all levels. However, the system can easily be changed and updated to meet new demands. For example, the current system automatically produces bulletins. This function could be developed further to produce automatically customized pest reports to include the interpretation of current pheromone or water traps data and to provide pertinent advice to growers and/or advisors. These could include comparisons between ear insect numbers in previous years and changes in the numbers between dates for specific regions, or more a forecast for weather conditions on the population dynamic of ear insects in any year.

The current expert system has improved understanding of the WBM problem. Improving risk prediction: the decision flow chart proposes thresholds to help predict the need for insecticide treatment. The verification study suggested that these are a good basis for risk management. However, thresholds are based on data from a limited number of sites and years; further work is required to confirm the initial findings and improve the precision with which it is possible to predict the risk of pest attack. Risk of damage is also primarily dependent upon the coincidence among midge activity, the wheat susceptible stage and weather conditions. Being able to predict the likely timing of the susceptible growth stage in relation to midge emergence would be a significant development, and help to limit unnecessary insecticide treatment.

WMTES integrates hypermedia advantages for information and knowledge supporting the ES recommendations for their better utility and increased dissemination capability through user friendly interfaces, as reported by Carrascal *et al.* (1995) for using hypermedia in knowledge transfer in agriculture. Estimation of pest activity using computational techniques has to incorporate variations from exact activity in the field. Practically, the relaxations given by the experts lead to acceptance in recommendations in up to 95% cases, except in the cases where differences in results are $\geq 10\%$. A knowledge base developed for a specific region in a problem can be tuned to make it more appropriate for other regions.

Much of the power and flexibility of expert systems are due to the fact that the knowledge base is separated from the inference mechanism (Waterman 1985; Rauscher 1990). The knowledge base can be modified without interfering with the operation of the system or the performance of other rules. In this way the program is now being enriched with rules for other insects, crops and sites, as well as by using of the user's feedback to improve the system. It is as important to develop the system as to maintain it. This task will be carried out for the extension services. WMTES is designed to enhance the quality and availability of knowledge required by decision makers in wheat insect management. It depends on a knowledge base that contains all the knowledge required to give useful, accurate and adequate consultations to wheat farmers.

CONCLUSION

The sequential sampling plans (pheromone traps, ear insect evaluation and water traps) described in this work should provide a method for more efficient midge monitoring. If pheromone trap catches indicate that a significant number of adults (30 midge males/ trap/ day) and suitable weather (temperature was >17°C, light wind and no rainfall) during the susceptible stage of the wheat crop, need to be closely monitored at growth stages 47-65 (Pivnick & Labbe 1993). Ear insect evaluation should be conducted in the milky stage (GS 73-75), when most larvae are already practically grown up, but have still not left the spike; the threshold is 3-4 larvae/ ear, as well >1 OWBM adults/ 6 wheat heads and >1 YWBM adults/ 3 ears at GS 55. The threshold of thrips was 10 thrips/ ear at heading stage and 25 thrips/ ear at flowering stage in case of companying with WBM, but this value was devised from scientific method. Water traps should be also monitored carefully with OWBM & YWBM after heavy rain, especially at late milky stage; the threshold was 40 L/ trap. This gives a reliable base for control decision making.

To minimize the economic and ecological impact of *S. mosellana*, wheat producers in Germany must be aware of management tools. Forecasts and risk warnings, monitoring tools, cultural control, agronomic practices, chemical control, biological control and plant resistance (so far, not completely approved) are all available for the industry to manage *S. mosellana*. Prior to the growing season, forecast maps predict high risk areas as mentioned by Olfert *et al.* (2004). If the rotation allows, the producer may choose not to grow wheat, grow a resistant variety of wheat as stated by Lamb *et al.* (2003), or grow an alternate resistant crop instead. If a lower degree of infestation is predicted, producers may stick to their plans to grow wheat, but may choose a less susceptible wheat variety and plant early to avoid high midge populations during heading as reported by Elliott *et al.* (2000). Wheat producers are urged to monitor crops closely in all areas where *S. mosellana* is present during the susceptible period (emergence of the wheat head from the boot until anthesis begins). If a lower degree of infestation is predicted, producers may stick to their plans to grow wheat, but may choose one of the resistant varieties or a less susceptible wheat variety and plant early to avoid high midge populations during heading. Using expert system for ear insects control is considered an example of small expert system technology can be successfully applied to the solution of every day problems in plant protection.

8. REFERENCES

Abdel-Aal E.S.M., Hucl P., Sosulski F.W., Graf R., Gillott C., Pietrzak L. (2001). Screening spring wheat for midge resistance in relation to ferulic acid content. *J Agric Food Chem* **49**: 3559-3566.

Agrisense™ (2007). "Orange Wheat Blossom Midge, Pheromone monitoring kit". Agrisense-BCS LTD. Treforest Industrial Estate, pontypridd, South Wales, CF37 5SU, UK.

Alford D. V. (1999). A Textbook of Agricultural Entomology. London, Blackwell Science Ltd.

Andjus L. (1996). The research into the Thrips fauna and significance of the plants of spontaneous flora for the survival of pest species. *Ph.D Thesis, Belgrade University*.

Andjus L. (2004). The thrips fauna on wheat and on plants of the spontaneous flora in the bordering belt surrounding it. *Acta Phytopathol et Entomol Hung* **39**: 255-261.

Anonymous (2008). List of authorized plant protection products in Germany with information on terminated authorizations. *Bundesamt für Verbraucherschutz und Lebensmittelsicherheit* 56, Auflage 2008. ISSN 0178-059X.

Anonymous (2009). Beschreibende Sortenliste Getreide, Mais, Ölfrüchte, Leguminosen (großkörnig), Hackfrüchte (außer Kartoffeln). *Bundessortenamt, Osterfelddamm 80, 30627 Hanover* ISSN 0948 – 4167, pp. 80-131.

Baker R.J. (1990) Agronomic performance of semi-dwarf and normal height spring wheats seeded at different dates. *Can J Plant Sci* **70**: 295-298.

Barker A.M., Sanbrooke K.J., Aebischer N.J. (1997). The water trap colour preferences of farmland sawflies. *Ento Exp Appl* **85**: 83–86.

Barnes H.F. (1928) Wheat blossom midges. Differences between *Contarinia tritici* and *Sitodiplosis mosellana*. *Bull Entomol Res* **18**: 285-288.

Barnes H.F. (1956). Gall Midges of Cereal Crops. Gall Midges of Economic Importance. London, Crosby, Lockwood & Son. Vol **VII**: 29-82.

Basedow T. (1971). "Zur morphologischen Unterscheidung der beiden Weizengallmückenarten *Contarinia tritici* (Kirby 1798) und *Sitodiplosis mosellana* (Géhin, 1857) (Dipt., Cecidomyidae). *Nachrichtenb Deutsch Pflanzenschutzd* **23**: 129-133.

Basedow Th. (1977). The susceptibility of some spring wheat varieties to the attack by the two wheat blossom midge species. *Anz Schäd Pflanzenschutz Umweltschutz* **50**: 129-131.

Basedow T. (1980). Untersuchungen zur Prognose des Auftretens der Weizengallmücken *Contarinia tritici* und *Sitodiplosis mosellana*. I. Die Kritischen Larvenzahlen im Boden. *Z Angew Entomol* **90**: 292-299.

Basedow T., Gillich H. (1982). Untersuchungen zur Prognose des Auftretens der Weizengallmücken *Contarinia tritici* und *Sitodiplosis mosellana*. II. Faktoren, die ein Schadauftreten der Mücken verhindern können. *Anz Schäd Pflanzenschutz* **55**: 84-89.

Bates B.A., Weiss M. J., McBride D. K. (1991). Biology and Management of Barley Thrips. North Dakota State University Extension Service, Fargo. *Pest Control and Pesticide Publications*; E1007.

Berzonsky W.A., Ding H., Haley S.D., Lamb R.J., McKenzie R.I.H., Ohm H.W., Patterson F.L., Peairs F.B., Porter D.R., Ratcliffe R.H., Shanower T.G. (2003). Breeding wheat for resistance to insects. *Plant Breeding Rev* **22**: 221–296.

Birkett M.A., Bruce T.J.A., Martin J.L., Smart L.E., Oakley J., Wahams L.J. (2004). Responses of female orange wheat blossom midge, *Sitodiplosis mosellana* to wheat panicle volatiles. *J Chem Ecol* **30**: 1319–1328.

Blake N.K., Stougaard R.N., Weaver D.K., Sherman J.D., Lanning S.P., Naruoka, Y. Xue Q., Martin J.M., Talbert L.E. (2010). Identification of a quantitative trait Locus for resistance to the orange wheat blossom midge in spring wheat. *Plant Breeding* (Accepted).

Brown M.B., Forsythe A.B. (1974). Robust Tests for Equality of Variances. *J Am Stat Assoc* **69**: 364–367.

Bruce T. J. A., Hooper A. M., Ireland L., Jones O. T., Martin J. L., Smart L. E., Oakley J., Wadhams L. J. (2007). "Development of a pheromone trap monitoring system for orange wheat blossom midge, *Sitodiplosis mosellana*, in the UK. *Pest Manag Sci* **63**: 49-56.

Carrascal M.J., Pau L.F., Reinet, L. (1995). Knowledge and information transfer in agriculture using hypermedia: a system review. *Comput& Electro in Agric* **12**: 83-119.

Cesevičienė J., Leistrumaitė A., Paplauskienė V. (2009). Grain yield and quality of winter wheat varieties in organic agriculture. *Agronomy Res* **7**: 217–223.

Chakraborty P., Chakrabarti D.K. (2008) A brief survey of computerized expert systems for crop protection being used in India. *Progress in Natural Science* **18**: 469-473.

Cuthbertson D. R. (1989). *Limothrips cerealium*: an alarming insect. *Entomologist* **108**: 246-256.

Ding H., Lamb R. J. (1999). Oviposition and larval establishment of *Sitodiplosis mosellana* on wheat (Graminae) at different growth stages. *Can Entomol* **131**: 475–481.

Ding H., Lamb R.J., Ames N. (2000). Inducible production of phenolic acids in wheat and antibiotic resistance to *Sitodiplosis mosellana*. *J Chem Ecol* **26**:969-985.

Doane J.F., Olfert O. (2008). Seasonal development of wheat midge, *Sitodiplosis mosellana*, in Saskatchewan, Canada. *Crop Prot* **27**: 951–958.

Doane J.F., Mukerji M.K., Olfert O. (2000). Sampling distribution and sequential sampling for subterranean stages of orange wheat blossom midge, *Sitodiplosis mosellana* (Géhin) in spring wheat. *Crop Prot* **19**: 427-434.

Edrees S.A., Rafea A., Fathy I., Yahia M. (2003). NEPER: a multiple strategy wheat expert system. *Comput & Electron in Agric* **40**: 27-43.

Elliott R.H. (1988). Evaluation of insecticides for protection of wheat against damage by the wheat midge, *Sitodiplosis mosellana* (Géhin) *Can Entomol* **120**: 615–626.

Elliott R.H., Mann L.M. (1996). Susceptibility of red spring wheat, *Triticum aestivum* L. cv. Katepwa, during heading and anthesis to damage by wheat midge, *Sitodiplosis mosellana* (Géhin) (Diptera: Cecidomyiidae). *Can Entomol* **128:** 367–375.

Elliott R.H., Mann L.M. (1997). Control of wheat midge, *Sitodiplosis mosellana* (Géhin), at lower chemical rates with small capacity sprayer nozzles. *Crop Prot* **16**: 235-242.

Elliott R.H., Mann L., Olfert O. (2000). Susceptibility of hard red spring wheats to damage by high populations of wheat midge. *SRC-Saskatoon Res Letter* 2000–09, 3 pp.

Ellis S.A., Bruce T.J.A., Smart L.E., Martin J.L., Snape J., Self, M. (2009). Integrated management strategies for varieties tolerant and susceptible to wheat midge. *HGCA Project Report number 451 May 2009*, 148 pages.

El-Wakeil NE, Gaafar N, Volkmar C (2010) Susceptibility of spring wheat to infestation with wheat midges and thrips. *J Plant Dis & Prot* **117**: 261–267.

El-Wakeil NE, Abdel-Moniem A, Gaafar N, Volkmar C (2013) Effectiveness of some insecticides on wheat blossom midges in winter wheat. Gesunde Pflanzen (Accepted).

ESICM (1994). A study of the needs assessment for expert systems in the agriculture sector in Egypt: Expert systems for improved crop management (UNDP/FAO, EGY/88/024). *Technical Report No. TR*-88-024-33.

Finch S. (1991). Influence of trap surface on the numbers of insects caught in water traps in brassica crops. *Ent Exp Appl* 59: 169-173.

Finch S. (1992). Improving the selectivity of water traps for monitoring populations of the cabbage root fly. *Ann Appl Biol* 120: 1–7.

Franssen C. J. H., Mantel W. (1965). Thrips in cereal crops (biology, economic importance and control). Versl. Landb. *Onderz Rijkslandb Proefstn* **662**: 97.

Freier B., Rossberg, D. (2001). Simulationsmodelle als Erkenntnismittel in der Agrarökologie. *IANUS* 1/2001.

Freier B., Pallutt B., Jahn, M., Sellmann, J., Gutsche, V., Zornbach, W., Moll, E. (2009). Network of referencefarms for plant protection. *Annual report of Julius Kühn-Institut 2008* **149**: 1-64.

Freier B., Triltsch H., Roßberg D. (1996). GTLAUS — a model of wheat–cereal aphid–predator interaction and its use in complex agroecological studies. *Z. Pflanzenkrankh. Pflanzenschutz* **103**: 543–554.

Freier B., Volkmar C., Lübke M., Wetzel T. (1982). Zur wirtschaftlichen Bedeutung der Ährenschädlinge im Getreidebau. Martin-Luther Univ. Halle Wittenb. Schaderreger in der Industriemässigen Getreideproduktion. *Wissensch Beitr* **37**: 116–127.

Gaafar N (2010) Wheat midges and thrips information system: Monitoring and decision making in central Germany. PhD Diss Martin- Luther- Uni Halle, 109 pages.

Gaafar N., Volkmar C. (2009). Monitoring system of orange wheat blossom midge, *Sitodiplosis mosellana* (Géhin) using pheromone trap in the central Germany. *Mitt Deutsch Gesell Allg und Angew Entomol* **17**: 221-225.

Gaafar N., Cöster H., Volkmar C. (2009). Evaluation of ear infestation by Thrips and wheat blossom midges in winter wheat cultivars. In: Feldmann F., Alford D.V., Furk C. (eds.) *Proc 3rd Internat Symp on Plant Prot & plant health in Europe, Berlin, Germany, 14-16 May 2009*, pp. 349-359.

Gaafar N., El-Wakeil N., Volkmar, C. (2011). Assessment of wheat ear insects in winter wheat varieties in central Germany. *J Pest Sci* **84**: 49-59.

Ganehiarachchi G.A.S.M., Harris M.O. (2007). "Oviposition behavior of orange wheat blossom midge on low- vs. high-ranked grass seed heads. *Ent Exp et Appl* **123**: 287-297.

Glen D.M. (2000). The effects of cultural measures on cereal pests and their role in integrated pest management. *Integ Pest Manag Rev* **5**: 25–40.

Gonzalez-Andujar J.L., Garcia de Ceca J.L., Fereres A., (1993). Cereal aphids expert system (CAES)" Identification and decision making. *Comput & Electron in Agric* **8**: 293-300.

Gonzalez-Diaz L., Martínez-Jimenez P., Bastida F., Gonzalez-Andujar J.L. (2009). Expert system for integrated plant protection in pepper (*Capsicum annuun* L.). *Expert Syst with Appl* **36**: 8975-8979.

Gosselke U., Triltsch H., Rossberg D., Freier B. (2001). GETLAUS01—the latest version of a model for simulating aphid population dynamics in dependence on antagonists in wheat. *Ecol Modell* **145**: 143–157.

Greene W. F. (2003). Econometric analysis (5th ed.). Pearson Education, *Upper Saddle River*, NJ.

Gries R., Gries G., Khaskin G., King S., Olfert O., Kaminski L.A., Lamb R. J., Bennett R. (2000). "Sex pheromone of orange wheat blossom midge, *Sitodiplosis mosellana*." *Naturwissenschaften* **87**: 450-454.

Harris M.O., Stuart J.J., Mohan M., Nair S., Lamb R.J., Rohfritsch O. (2003). Grasses and gallmidges: Plant defense and insect adaptation. *Ann Rev Entomol* **48**: 549–577.

Helenius J., Kurppa S. (1989). Quality losses in wheat caused by orange wheat blossom midge *Sitodiplosis mosellana*. *Ann Appl Biol* **114**: 409-417.

Hübner M., Wittkopf R. (2010). Pflanzenschutzempfehlung des amtlichen Pflanzenschutzdienstes. Ackerbau& Grünland, *LLFG Jahresbericht* (pp. 47) www.llfg.sachsen-anhalt.de.

Hurvich C.M., Tsai C.L. (1989). Regression and time series model selection in small samples. *Biometrika* **76**: 297-397.

Jörg E., Racca P., Preiß U., Buttutini A., *et al.* (2007). Control of Colorado potato beetle with the SIMLEP decision support system. *OEPP/EPPO* **37**: 353-358.

Kakol E., Kucharczyk H. (2004). The occurrence of Thrips (Thysanoptera) on winter and spring wheat in chosen regions of Poland. *Acta Phytopathol et Entomol Hung* **39**: 263-269.

Khan F.S., Razzaq S., Irfan K., Maqbool F., Farid A., Illahi I., Ul amin T. (2008). Dr. Wheat: a web-based expert system for diagnosis of diseases and pests in Pakistani wheat. World Congress on Engineering, London, July 2 - 4, 2008. *Proc of WCE* **I**: 549-554.

Kleinhenz B. (2007). Neue Prognosen aus dem Internet." *DLG-Mitteilungen* **9**: 42-43.

Kleinhenz B., Roßberg D. (2000). Structure and development of decision-support systems and their use by the State Plant Protection Services in Germany. *Bull OEPP/EPPO* **30**: 93-97.

Kleinhenz B., Zeuner T. (2007). Introduction of GIS in decision support systems for plant protection. *In* Alord D.V., Feldman F., Hasler J., Tiedemann A. *Proc 2rd Internat Symp on Plant Prot & plant health in Europe,* Humboldt Universität, Berlin, 2007, **82**: 24-25.

Knight J. D., Tatchell G. M., Norton G. A., Harrington R. (1992). FLYPAST: an information management system for the Rothamsted Aphid Database to aid pest control research and advice. *Crop Prot* **11**: 419-426.

Knodel J. (1995). "Orange Wheat Blossom Midge *Sitodiplosis mosellana*", [online]. North Dakota State University, Fargo North Dakota [16.01.2008] 22.02.2007. http://www.ag.ndsu.nodak.edu/aginfo/entomology/entupdates/Wheat_Midge/owbm.htm

Köppä P. (1969). Studies on the hibernation of certain species of thrips living on cereal plants. *Ann Agric Fenn* **8**: 1–8.

Köppä P. (1970). Studies on the thrips species most commonly occurring on cereals in Finland. *Ann Agric Fenn* **9**: 191–265.

Kucharzyk H. (1998). Thysanoptera and other insects collected in differently coloured traps in eastern Poland. *Proc 6th Internat Symp on Thysanoptera*, pp, 81-87.

Kurppa S. (1989). "Wheat blossom midges, *Sitodiplosis mosellana* (Gehin) and *Contarinia tritici* (Kirby) in Finland during 1981-87." *Ann Agric Fenn* **28**: 87-96.

Lamb R.J., Wise I.L., Olfert O.O., Gavloski J., Barker P. S. (1999). "Distribution and seasonal abundance of *Sitodiplosis mosellana* in spring wheat. *Can Entomol* **131**: 387-397.

Lamb R.J., McKenzie R.I.H., et al. (2000a). "Resistance to *Sitodiplosis mosellana* (Diptera: Cecidomyiidae) in spring wheat (Gramineae). *Can Entomol* **132**: 591-605.

Lamb R.J., Tucker J.R., Wise I.L., Smith, M.A.H. (2000b). Trophic interaction between *Sitodiplosis mosellana* (Diptera) and spring wheat: implications for yield and seed quality. *Can Entomol* **132**: 607–625.

Lamb R.J., Smith M.H., Wise I.L., Clarke P., Clarke J. (2001). Oviposition deterrence to *Sitodiplosis mosellana*: A source of resistance for wheat. *Can Entomol* **133**: 579- 591.

Lamb R.J., Wise I.L., Smith M. H., McKenzie R.I.H., Thomas J., Olfert O.O. (2002) Oviposition deterrence against *Sitodiplosis mosellana* in spring wheat. *Can Entomol* **134**: 85- 96.

Lamb R.J., Sridhar P., Smith M.A.H., Wise I.L. (2003). Oviposition preference and offspring performance of a wheat midge *Sitodiplosis mosellana* on defended and less well defended wheat plants. *Environ Entomol* **32**: 414–420.

Larsson H. (1988). Economic damage caused by cereal thrips in winter rye in Sweden. *Acta Phytopathol Entomol Hung* **23**: 291–293.

Larsson H. (2005). Aphids and Thrips: Dynamics and Bio-Economics of Cereal Pests. *Doctoral thesis Swedish University of Agricultural Sciences*, 42 pages.

Lattauschke G., Wetzel T. (1985). Zum Artenspektrum und zur Abundanzdynamik von Getreide-Thysanopteren. *Arch Phytopathol Pflanzenschutz* **21**: 375-382.

Lattauschke G., Wetzel T. (1986). Zum Auftreten und zur Bedeutung von Thysanopteren in Getreidebau. *Nachrichtenbl Pflanzenschutz DDR* **40**: 162–165.

Levene H. (1960). Robust tests for the equality of variance. Contributions to probability and statistics; Essays in Honor of Harold Hotelling. I. Olkin *et al.* (eds.) *Stanford Uni Press*, pp. 278-292.

Lewis T. (1973). Thrips, their biology, ecology and economic importance. *Academic Press* London and New York.

Lucus P. (1997). Symbolic diagnosis and its formalization, Knowle. *Eng Rev* **12**: 109–146.

Mann B.P., Wratten, S.D., Watt, A.D., (1986). A computer-based advisory system for cereal aphid control. *Comput & Electron Agric* **1**: 263-270.

McKenzie R.I.H., Lamb R.J., *et al.* (2002). "Inheritance of resistance to wheat midge, *Sitodiplosis mosellana*, in spring wheat. *Plant Breeding* **121**: 383-388.

Meers S. (2004). Barley Thrips. Economic importance. Agriculture, food and rural Development. Alberta Government. www.agric.gov.ab.ca/app21/seltopcat?cat1= Diseases/Insects/Pests.

MölcK G. (2006). Erfahrungen mit Prognose und Bekämpfung des Schadauftretens von Sattelmücken Weizengallmücken (Diptera: Cecidomyiidae) in Schleswig-Holstein. *Mitt. BBA* **400**: 227.

Moritz G. (2006). Die Thripse. *Die Neue Brehm- Bücherei Bd. 663*, pp, 384.

Mound L. A. (1971). The feeding apparatus of thrips. *Bull Ent Res* **60**: 547–548.

Mound L. A. (1997). Biological diversity. *Thrips as crop Pests*. Lewis T. (ed) pp.107-216. CAB International, Wallingford, UK.

Mound L. A. (2005). Thysanoptera: Diversity and Interactions. *Ann Rev Entomol* **50**: 247-269.

Oakley J N. (1994). Orange Wheat Blossom Midge: A Literature Review and Survey of the 1993 Outbreak. *Research Review* No. 28, HGCA, London.

Oakley J.N. (2008). Control needs for changing pest distribution. Arable cropping in a changing climate: *Proc HGCA Conf United Kingdom: HGCA*. Pp. 87–92.

Oakley J.N., Ellis S.A., Walters K.F.A., Watling M. (1993). The effects of cereal aphid feeding on wheat quality. *Aspects of Applied Biology 36, Cereal Quality III*, pp. 383-390.

Oakley J.N., Cumbleton P.C., Corbett S.J., Saunders P., Green D.I., *et al.* (1998). "Prediction of orange wheat blossom midge activity and risk of damage. *Crop Prot* **17**: 145-149.

Oakley J.N., Talbot G., Dyer C., Self M.M., Freer J.B.S. *et al.* (2005). Integrated control of wheat blossom midge: variety choice, use of pheromone traps and treatment thresholds." HGCA publications, London. *HGCA Project Report* (363): 65.

O'Brien RG (1981) A simple test for variance effects in experimental designs. *Psychol Bull* **89**: 570-574.

Olfert O., Elliott, R.H., Hartley, S. (2009). Non-native insects in agriculture: strategies to manage the economic and environmental impact of wheat midge, *Sitodiplosis mosellana*, in Saskatchewan. *Biol Invasions* **11**: 127–133.

Olfert O., Elliott R.H., Hartley S., Zeleny K. (2004) Forecast of wheat midge in Alberta and Saskatchewan for 2004. 2003 Crop variety highlights and insect pest forecasts. *Saskatoon Res Centre Technical Bull No.* 2004–01, pp: 14–15.

Olfert O.O., Mukerji M.K., Doane J.F. (1985). Relationship between infestation levels and yield loss caused by wheat midge, *Sitodiplosis mosellana* (Géhin) (Diptera: Cecidomyiidae), in spring wheat in Saskatchewan. *Can Entomol* **117**: 593-598.

Parrella M.P., Lewis T. (1997). Integrated pest management in field crops. Thrips as crop Pests Lewis T. (ed.) pp. 595-614. *CAB International*, Wallingford, UK.

Petersen G., Heimbach U. (2009) Ergebnisprotokoll der 19. Tagung des DPG-Arbeitskreises Integrierter Pflanzenschutz, Arbeitsgruppe „Schädlinge in Getreide und Mais". *J Kulturpflanzen* **61**: 178-179.

Pivnick K. A. (1993). "Response of males to female sex pheromone in the orange wheat blossom midge, *Sitodiplosis mosellana* (Gehin) (Cecidomyiidae)." *J Chem Ecol* **19**: 1677-1689.

Pivnick K.A., Labbe E. (1992). "Emergence and calling rhythms, and mating behaviour of the orange wheat blossom midge, *Sitodiplosis mosellana. Can Entomol* **124**: 501-507.

Pivnick K.A., Labbe E. (1993). "Daily patterns of activity of females of the orange wheat blossom midge, *Sitodiplosis mosellana* (Gehin).*Can Entomol* **125**: 725-736.

Räder T., Racca P., Jörg E., Hau, B. (2007). PUCREC/PUCTRI - a decision support system for the control of leaf rust of winter wheat and winter rye. *OEPP/EPPO* **37**: 378-382.

Rafea A., Shaalan K. (1996). Using expert systems as a training tool in the agriculture sector in Egypt. *Expert Syst with Appl* **11**: 343-349.

Rafea A., El-Dessouki A., Hassan H., Mohamed S. (1993). Development and implementation of a knowledge acquisition methodology for crop management expert systems. *Comput & Electron in Agric* **8**: 129-146.

Rauscher H.M. (1990). Practical expert system development in Prolog. *Artif Intell Appl Natur Resour Manage* **4**: 51-55.

Rieckmann W., Block T., Frosch M., Heimbach U., Lein K.-Lauenstein G., Matthes P., Steck U., Volkmar C. (2001). Deutscher Vorschlag für eine EPPO-Richtlinie zur Prüfung von Insektiziden gegen Gallmücken an Getreide. *BBA* **1**.35.

Saini H.S., Kamal R., Sharma A.N. (2002). Web Based Fuzzy Expert System for Integrated Pest Management in Soybean. *Inter J Inf Technol* **8**: 54-74.

SAS (2009) Institute Inc., *Cary, NC*. SAS software.

Schleich-Saidfar C, Gleser HJ, Petersen G. (2007) Schädlinge in Winterweizen. Ämter-für-ländliche-Räume Versuchsbericht Ackerbau. Husum, Kiel & Luebeck, Pflanzenschutzdienst des Landes Schleswig-Holstein.

Schröder A. (2009) Evaluierung von Winterweizensorten zum Auftreten und Schadausmaß von Thysanopteren. Diplomarbeit Martin- Luther- Universität Halle- Wittenberg. 94 pages.

Seidel D., Wetzel T., Bochow H. (1983). Pflanzenschutz in der Pflanzenproduktion. Berlin: *VEB Deutscher Landwirtschaftsverlag*, pp. 79–80.

Sharma M.C. (2001). Integrated pest management in developing countries: Special Reference to India, *IPM Mitr*, pp. 37-44.

Sivakami S., Karthikeyan C. (2009). Evaluating the effectiveness of expert system for performing agricultural extension services in India. *Expert Syst with App* **36**: 9634-9636.

Smith M.A.H., Lamb R.J. (2001). "Factors influencing oviposition by *Sitodiplosis mosellana* (Diptera: Cecidomyiidae) on wheat spikes (Gramineae). *Can Entomol* **133**: 533-548.

Smith M.A.H., Lamb R.J. (2004). "Causes of variation in body size and consequences for the life history of *Sitodiplosis mosellana*. *Can Entomol* **136**: 839-850.

Smith M.A.H., Wise I.L., Lamb R.J. (2004). "Sex ratios of *Sitodiplosis mosellana* (Diptera): implications for pest management in wheat. *Bull Entomol Res* **94**: 569-575.

Smith M.H., Wise I.L., Lamb, R.J. (2007). Survival of *Sitodiplosis mosellana* on wheat with antibiosis resistance: implication for the evolution of virulence. *Can Ent* **139**: 133–140.

Suo X., Shi N. (2008). Web-based expert system of wheat and corn growth management. IFIP *Internat Federation for Inf Process* **258**: 111-119.

Tanskii V.I. (1961). Formation of thrips (Thysanoptera) fauna of wheat sowings on new lands of North Kazakhstan. *Entomol Obozr* **40**: 785-793.

Thomas C.R., Maurice S.C. (2008). Statistix 9. *Managerial Economics McGraw-Hill/Irwin*, (ISBN: 0073402818), More information at http://www.statistix.com.

Thomas J.B., Fineberg N., Penner G.A., *et al.* (2005). Chromosome location and markers of Sm1: a gene of wheat that conditions antibiotic resistance to orange wheat blossom midge. *Molecular Breeding* **15**: 183-192.

Tiedemann A., Kleinhenz B. (2008). Prognose contra Praxis? *DLG-Mitteilungen* **3**: 54-58.

Tottman D.R., Broad H. (1987). Decimal code for the growth stages of cereals. *Ann Appl Biol* **110**: 683–687.

USDA (2007). Annual World Production Summary, Grains, http://www.usda.gov/, retrieved on 4 August 2010.

Volkmar C. (1988). Zum kombinierten Auftreten und zur Bekämpfung von Ährenschädlingen an Winterweizen unter Praxisbedingungen. *Tag-Ber, Akad Landwirtsch-Wiss, Berlin DDR* **271**: 577-580.

Volkmar, C. Wetzel, T. (1989) Zum Auftreten und zur Bekämpfung von Ährenschädlingen des Winterweizen unter Praxisbedingungen. *Nachrichtenbl Pflanzenschutz DDR* 42:14-17.

Volkmar C., Werner C., Matthes P. (2008). On the occurrence and crop damage of wheat blossom midges *Contarinia tritici* (Kby.) and *Sitodiplosis mosellana* (Geh.) in Saxony-Anhalt. *Mitt Dtsch Ges Allg Angew Ent* **16**: 305-308.

Volkmar C., Schröder A., Gaafar N., Cöster H., Spilke, J. (2009). Evaluierungsstudie zur Befallssituation von Thripsen in einem Winterweizensortiment. *Mitt Dtsch Ges Allg Angew Ent* **17**: 227-230.

Volkmar, C. Gaafar, N. (2010) Kontrolle mit Pheromonfallen. *DLG- Miteillung* 4: 64-67

Waterman D.A. (1985). A Guide to Expert Systems. *Addison-Wesley, Reading, MA*, 441 pp.

Wetzel T. (1964). Untersuchungen zum Auftreten, zur Schadwirkung und zur Bekämpfung von Thysanopteren in Grassamenbeständen. *Beitr Entomol* **14**: 427–550.

Wetzel, Th., Freier, B. (1981). Bekämpfungsrichtwerte für Schädlinge des Getreides. *Nachrichtenbl Pflanzenschutz DDR* **35**: 47-50.

10. Photo gallery

Pheromone traps- Monitoring method

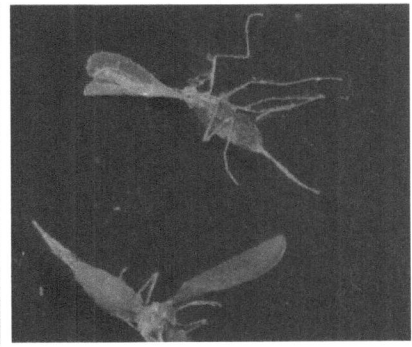
Differences between orange & yellow wheat midge female

External damage symptom of WBM

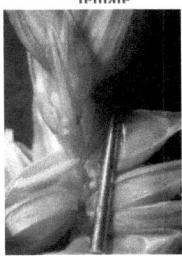
Yellow wheat midge damage

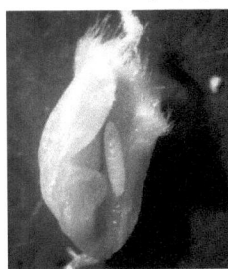
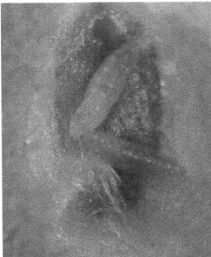
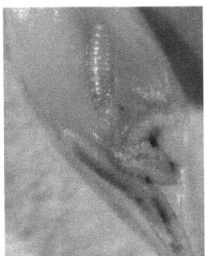
Different infestation level of orange wheat midge damage

Normal and damaged kernels

OWBM damage in resistant variety   OWBM damage in susceptibl variety

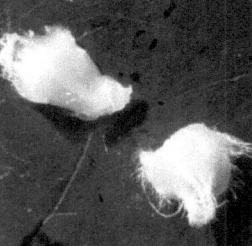
WBM damage after harvest time

10. Index

1. SUMMARY..3

2. ZUSAMMENFASSUNG..6

3. INTRODUCTION...9

4. SCIENTIFIC BACKGROUND..11
4.1. Taxonomy, morphology, biology and economic significance of wheat midges.........11
4.2. Taxonomy, morphology, biology and economic significance of wheat thrips...........15
4.3. Expert system...16

5. MATERIAL AND METHODS..17
5.1. Monitoring sites..17
5.1.1. Silstedt site...17
5.1.2. Halle site (winter and spring wheat) 20
5.1.3. Salzmünde site...25
5.2. Monitoring methods...25
5.2.1. Survey *S. mosellana* adults using pheromone traps......................................25
5.2.2. Evaluation of thrips and midges in wheat ears...26
5.2.3. Inspecting wheat midge larvae using white water traps...................................26
5.3. Expert system verification...27

6. RESULTS...28
6.1. Winter wheat varieties in Silstedt..28
6.1.1. Population of *S. mosellana* adult surveyed using pheromone traps..............28
6.1.2. Evaluation of thrips and midges in wheat ears...28
6.1.3. Wheat midge larvae population inspected using water traps...........................37
6.1.4. Categorizing of wheat varieties based on ear insect infestation 39
6.2. Winter wheat varieties in Halle...40
6.2.1. Population of *S. mosellana* adult surveyed using pheromone traps..............40
6.2.2. Evaluation of thrips and midges in wheat ears...41

6.2.3. Wheat midge larvae population inspected using water traps...........................47
6.2.4. Groupings of wheat varieties based on infestation levels..............................49
6.3. Comparison between ear insect infestation levels 2008...........................51
6.3.1. Halle site...51
6.3.2. Silstedt site..53
6.3.3. Comparison between Halle and Silstedt..55
6.4. Spring wheat varieties in Halle..57
6.4.1. Population of *S. mosellana* adult surveyed using pheromone traps.......57
6.4.2. Evaluation of thrips and midges in wheat ears..57
6.4.3. Wheat midge larvae population inspected using water traps..................61
6.4.4. Spring wheat yield...64
6.5. Winter wheat fields in Salzmünde..67
6.5.1. Population of *S. mosellana* adult surveyed using pheromone traps.......67
6.5.2. Evaluation of thrips and midges in wheat ears..69
6.5.3. Wheat midge larvae population inspected using water traps..................73
6.6. Wheat Midges and Thrips Expert System (WMTES)...............................75
6.6.1. General information..75
6.6.2. Model Verification Study (Methodology) 77
6.6.3. Expert System Development Life Cycle and Structure............................78
6.6.4. Testing and validation...89
6.6.5. What should be concluded?..89
7. DISCUSSION..92
8. REFERENCES...102
9. PHOTO GALLERY..112
10. Index 113

i want morebooks!

Buy your books fast and straightforward online - at one of world's fastest growing online book stores! Environmentally sound due to Print-on-Demand technologies.

Buy your books online at
www.get-morebooks.com

Kaufen Sie Ihre Bücher schnell und unkompliziert online – auf einer der am schnellsten wachsenden Buchhandelsplattformen weltweit! Dank Print-On-Demand umwelt- und ressourcenschonend produziert.

Bücher schneller online kaufen
www.morebooks.de

 VDM Verlagsservicegesellschaft mbH
Heinrich-Böcking-Str. 6-8   Telefon: +49 681 3720 174   info@vdm-vsg.de
D - 66121 Saarbrücken   Telefax: +49 681 3720 1749 www.vdm-vsg.de

Printed by Books on Demand GmbH, Norderstedt / Germany